ADMINISTRATIVE PRACTICES IN CONCRETE CONSTRUCTION

ADMINISTRATIVE PRACTICES IN CONCRETE CONSTRUCTION

PORTLAND CEMENT ASSOCIATION

JOHN WILEY & SONS, INC.

NEW YORK LONDON SYDNEY TORONTO

Library of Congress Cataloging in Publication Data

Portland Cement Association.
Administrative practices in concrete construction.

"Fourth in a series of five textbooks that comprise the
National concrete technology curriculum."
Includes index.
1. Concrete construction–Contracts and specifications–United States. 2. Construction industry–Law and Legislation–United States. I. Title.
KF1950.P6 343'.73'078 74-28259
ISBN 0-471-67433-8

Printed in the United States of America

10 9 8 7 6 5 4 3 2 1

ACKNOWLEDGMENTS

This series is a national program to educate persons for employment in the concrete industries, sponsored by the Portland Cement Association, the National Ready Mixed Concrete Association and the American Concrete Institute in cooperation with the Office of Education, U.S. Department of Health, Education, and Welfare.

PROJECT STAFF

George R. White,
Project Director
J. David Jackson,
Senior Curriculum Editor
David Anderson,
Associate Editor
William Perenchio
Technical Advisor
Elizabeth A. Craig,
Administrative Assistant
Cynthia A. Holck,
Editorial Assistant

TECHNICAL REVIEWERS

The staff acknowledges the many excellent contributions by outstanding people in the concrete field and the advisory group that made up the administrative team. Of special significance were contributions by

Howard Wiechman, John Seeger, John Metcalf, Dr. J. R. D. Brown, and others who helped launch this undertaking. Special mention goes to the instructors and students who tested the trial materials at the pilot schools and whose comments and suggestions have been incorporated in these volumes.

DISCLAIMER STATEMENT

This work was developed under a grant from the U.S. Office of Education, Department of Health, Education and Welfare. However, the opinions and other content do not necessarily reflect the position of the agency, and no official endorsement should be inferred.

This publication is based on the facts, tests, and authorities stated herein. It is intended for the use of professional personnel who are competent to evaluate the significance and limitations of the reported findings and who will accept responsibility for the application of the material that it contains. The Portland Cement Association and publication cosponsors disclaim any and all responsibility for the application of the stated principles or for the accuracy of any of the sources other than work performed or information developed by the Association.

Caution: Avoid prolonged contact between unhardened (wet) cement or concrete mixtures and skin surfaces. To prevent such contact, it is advisable to wear protective clothing. Skin areas that have been exposed to wet cement or concrete, either directly or through saturated clothing, should be thoroughly washed with water.

PREFACE

A concrete technologist who is employed by a large company may have little contact with administration. Nevertheless, the more he knows about the business of concrete, the more valuable he will be to his employer. And the technologist who works for a small company may even be assigned administrative responsibilities.

This book provides a survey of three areas of administrative responsibility: legal aspects, planning and estimating, and compilation and evaluation of data.

A person involved in the construction business accepts certain legal obligations to the parties with whom he contracts and also to the community in which the work will be done. Furthermore, he has legal obligations to the suppliers of his materials and to his employees. Civil and criminal law regulate these relationships to ensure that fair and just expectations are enforced. The technologist need not be a lawyer, but if he is going to function smoothly and successfully, he must have a working knowledge of the applicable laws.

A frequent cause of failure in the construction business is faulty estimating and planning. What is estimating? It is predicting the amounts of materials, manpower, and equipment that will be required for a job and multiplying each by its current cost. This job can be futile, however, if there is not adequate planning and scheduling for job completion on time. Estimating is an essential part of all construction work, and it is necessary in bidding for contracts, in the planning and scheduling of jobs, and in operating the construction company at a profit.

The term "statistical data" is often confusing to the layman, but the effective technologist should have at least an elementary comprehension of the subject. The final section of the book deals with the collection, organization, and presentation of data, characteristics of frequency distribution, and various limits.

Administrative Practices in Concrete Construction is the fourth in a series of five books that comprise the National Concrete Technology Curriculum. The U.S. Office of Education funded the preparation of these instructional materials to assist in meeting the construction industry's continuing need for the development of personnel with a fundamental knowledge of concrete technology. The first three books in the series are *Principles of Quality Concrete, Basic Concrete Construction Practices*, and *Special Concretes and Concrete Products*. The text that follows this one is *Concrete Inspection Procedures*. The Series is supplemented by a *Laboratory and Exercise Manual on Concrete Construction* and *Instructor's Guides*.

PORTLAND CEMENT ASSOCIATION

TABLE OF CONTENTS

ADMINISTRATIVE PRACTICES IN CONCRETE CONSTRUCTION

SECTION ONE
LEGAL ASPECTS

CHAPTER 1

LEGAL ASPECTS OF CONCRETE CONSTRUCTION

At the beginning of any concrete construction project the originator has an idea of the final form of the project. Such an idea affects no one in the community. However, once work begins on the project (from paperwork to final construction), certain interrelationships and obligations are established with the community as well as with individuals. Such interrelationships involve duties and rights that must be observed between persons and groups. Nonobservance of these rights and duties is the basis of the legal concept of "liability."

The interrelationships involved in the execution of a concrete project are more numerous than would first appear. The contractual arrangements or agreements, whether written or unwritten, established between the originator (the owner) and those to be entrusted with the final execution (the architect, engineer, contractor) are important. But, so too are the arrangements made with other persons. Included within this classification are the workmen and the suppliers of materials as well as the subcontractors and members of other specialized trades. Many of the questions that arise during the field execution of a concrete construction project are matters of interpretation of such agreements. As such, they fall within that category known as "contract law."

In many instances, even though the actions of the parties involved are within the scope of existing agreements, questions arise. For example, are the provisions of an agreement fulfilled if a substitute accomplishes the designated tasks instead of the scheduled person? Such a situation would be in the category of the law known as the "law of agency."

In incidents of personal injury, whether they involve a workman on the construction project or an innocent bystander, reference would be made to the principles of the "law of torts."

Distinctions must be made as to whether the violation of rights affect an individual or the community as a whole. If the violation would be injurious to the entire community, it might fall within the realm of "criminal law." On the other hand, governmental regulation might be such that an administrative agency would have the right to deal with the matter. Such incidents would be controlled by the Workmen's Compensation Act.

There are also certain legal requirements to be met in the financing of a project that are within the scope of the law of negotiable instruments. Ways also exist to minimize one's liability by the use of insurance and surety bonds.

The legal aspects of the construction field are so varied that they are hard to keep track of and to cover sufficiently. However, the leading questions that must be asked to give any possible situation its proper perspective are:

1. Has a wrong occurred that would give rise to a legal liability?
2. If so, who is legally liable?
3. In what way can the situation be remedied?

WHAT IS THE GENERAL NATURE OF LEGAL LIABILITY?

Laws are the guidelines for the manner in which we exist within our society. As such, they are merely a collection of rules that has grown in number, refinement, and complexity. The laws that control an individual's actions within his community are known as the "rules of civil conduct."

All rules of civil conduct are concerned with rights, wrongs, and duties. The rights of one person automatically require duties on the part of another person to observe such rights. A wrong occurs whenever there is a violation of either a right or a duty. The person committing such a wrong is therefore responsible for any damage occasioned to another person's physical well-being or property.

Basic, clear-cut situations are rarely to be found in actual circumstances. Those situations requiring legal advice usually require decisions as to whether a wrong has been committed or the identity of the party or parties committing the wrong. Simply stated, it is a matter of fixing the liability for wrongful acts, if indeed any have occurred.

Liability, in the sense in which it is used in the law courts, is an obligation that one party owes to another party. This obligation may take the form of the payment of money, or the form of project completion in accordance with the terms of an executory contract. An executory contract is one on which no action has been taken, and the requirement of such completion would be termed "specific performance." One of the types of action that a person may be required to take by law can be deemed to be a "negative" action—that is, refraining from doing something. An example of this would be an injunction prohibiting proceeding with construction without first completing certain safety precautions to safeguard the general public. Or the specific action required may be to repair or restore something that has been damaged by the person to whom the liability has attached. For example, if in a construction project, a contractor damages a wall belonging to the neighboring property owner, the contractor may be required to repair the wall.

In real-life situations it is the rare incident where all liabilities and remedial measures are so well clarified as to present no problems or questions. To illustrate the way in which problems and questions can arise, let us refer to a recent court action that took place in New Jersey.

In this instance, the construction project involved the laying of a new sewer line along an avenue. In the course of the work, the sides of the trench in which the line was being laid collapsed and two workmen were killed. These facts would seem to present a very simple problem concerning liability for the deaths of the two men. They were the contractor's workmen engaged in the work for which they were paid—a clear case of workmen's compensation.

However, let us go a little more deeply into the facts. The contract covering the construction project involved was between the village and the contractor. The plans covering the projected line of the sewer were provided by the village and on the plans there were indications of a water main extending in the same direction as the proposed sewer line at a distance of some 12 feet from the sewer line itself. As the work progressed, however, it was discovered that instead of being 12 feet away from the construction area, in some instances the water main lay adjacent to the trench being dug for the sewer line. It was also discovered that at several places the joints in the water main were leaking. The contractor registered a complaint with the village that the presence of the leaking water main constituted a hazardous condition for his workmen. The village did take some steps to correct the condition. We can see now that there is a question concerning the adequacy of the measures taken to protect the workmen, and that there is an additional problem as to whether the village or the contractor should take responsibility.

The matter was brought into court by the administrators of the estates of the two victims. The defendant in the case was the village. Those bringing the suit, charged that the village was negligent in not safeguarding the workmen. Inasmuch as the workmen were engaged on a project for the village's benefit, it was the duty of the village to protect them from injury.

It is easy to understand what the village answered in response to the plaintiff's claims. The village maintained that the men were the contractor's employees and that it was the contractor's responsibility to provide sufficient support for the sides of the excavation and to remove any water.

It must be recalled that rules and regulations vary from locality to locality. This is also true of court decisions. The matters that were to be decided were:

Did the village owe a duty to the deceased workmen?

If such a duty existed, did the village breach its duty?

If such a breach of duty occurred, was the breach the cause of the workmen's deaths?

In answer to the contention of the village that the contractor was responsible, the court held that the duty of the contractor did not extend beyond such support of the excavation sides and removal of water that would be "normal and incidental" to the contract. The repair and correction of conditions resulting from a leaking water main, not a part of the contract, was not normal and incidental to the contract. Therefore, the village was liable if a breach of duty occurred.*

Rights, Wrongs, and Duties

The terms "rights," "wrongs," and "duties," the very basis of legal liability, can have many shades of meaning. Further exploration shows how the body of law has been built on the concept that each individual is

* Zanca *v.* Conti, 72 N. J. Super 23, cited in *Engineers & Contractors Magazine,* April 1969.

obliged to observe certain rules and regulations with respect to all other individuals.

General Considerations

As individuals, almost every major event in our lives has attached to it a significant legal document—birth certificate, marriage certificate, army discharge, death certificate. In our working careers, we have contracts, promissory notes, mortgages, loan agreements, and licenses. Licenses are assuming additional importance as evidence that one is qualified to enter into business relationships with others or participate in a profession or industrial specialization.

Such relationships with the public at large are also governed by legal considerations that may be either written or unwritten, formal in character or understood to exist through one's actions.

Consider a recent case that serves to illustrate the interrelationship that can exist for a law as it appears on the statutes of state and the results of one's own actions.

The mechanics Lien Act of the State of New Jersey provides that if the owner of property on which construction is to take place files a copy of the construction contract in the office of the county clerk, the rights to a lien on the property by a third party to the contract are extinguished. Let us clarify this by pointing out how such a lien by a third party can arise.

A lien is a charge made against either real estate or personal property for the payment of a monetary obligation. Once a lien has been placed against a piece of property, the owner is prevented from selling or otherwise disposing of the property without paying off the lien. Before he can sell, he must satisfy the lien, as well as any interest that may have been added.

In a construction contract, the contracting parties are usually the owner of the real property and the contractor. Ordinarily, the contractor agrees to furnish such materials and labor as will be needed for the completion of the project covered by the contract. If the contractor fails to pay his workmen's wages or to pay for materials received from various manufacturers, in many instances the workmen or suppliers have placed liens against the property that has received the benefits of their services or materials. Liens are registered with the Recorder or County Clerk of the locality in which the property is located.

However, it can be seen that under the statute mentioned earlier, the owner of the property on which construction is to commence can protect himself from a lien of this type by filing a copy of the construction contract with the County Clerk. In a way, it can be said that in New Jersey, the owner puts all other persons on notice that he will not be responsible for any bills of the contractor by filing a copy of the contract with the County Clerk.

There is an additional provision in the law under consideration that should the contractor fail to pay such bills, those to whom the money is due can file stop notices with the County Clerk. The owner then must hold out the amount of money covered by such stop notices from the money due the contractor under the contract and pay the withheld money to those filing the stop notices. It might be well to mentally distinguish the differences between such stop notices and the right to a lien.

In one case a suit was brought into court by a company that had supplied materials to a contractor. The contractor had failed to pay for these items. Since a copy of the construction contract had been filed by the owner, the supplier filed a stop notice. The owner, however, refused to honor this stop notice, and the supplier brought suit against the owner for payment of the money owed.

Under the facts of the case, the construction contract between the owner and the contractor required that an architect's certification be provided before the owner would be obligated to pay. Although the construction had been completed, the contractor had not provided such an architect's certification. The owner claimed that as he did not owe the contractor any money because of failure on the part of the contractor to fulfill the provision leading to payment, he could not be held liable for the bills of the supplier. In effect, he maintained that any money he would pay the supplier would have to be withheld from money owed the contractor, and he did not owe the contractor any money.

The court held that as the owner had received benefits under the contract—that is, he did have a completed construction project. Such conduct on his part in accepting the construction constituted a waiver of the provision requiring the architect's certification before payment would be made. The owner, therefore, owed money to the contractor. Whether the amount owed would be the contract price or a price adjusted because of the absence of the architect's certification is immaterial since the owner did owe money to the contractor. Thus there were funds that could be withheld to pay the supplier the amount covered by the stop notice.*

* Sargent Bros., Inc. *v.* Brancati, Ct. of E. & A. of N. J., 107 N. J. L. 84, cited in *Engineers & Contractors Magazine,* April 1969.

The body of the laws under which we live has developed from decisions about what is fair and just. Such decisions have in turn become codified as "rules of civil conduct" and the decisions constitute the guidelines that govern our everyday existence.

There is a written embodiment of rules and regulations with which one should become familiar: the Municipal Building Code. In the case of the Municipal Building Code, the rules that govern a building project are those which are established by the government. However, should there be additional requirements set forth by the state regarding projects, the rules of the state will prevail. The federal government takes precedence over the state.

CLASSIFICATION OF WRONGS

A wrong may be classified as belonging to one of two types, criminal or civil. A criminal wrong is more commonly known as a "crime," or a violation of the rights of the public at large. In this definition, a crime is not only homicide and burglary, but such things as reckless driving and disorderly conduct as well.

One charged with a crime is, therefore, prosecuted by the people. The title of a legal case sets forth clearly whose rights have been violated and who is being accused of committing the violation. The one who claims his rights have been violated and who brings the suit to the courts is known as the "plaintiff." The plaintiff's name is the first name in the title of a case. The one accused of the violation is known as the "defendant." The defendant's name is the second name in the title of the case. The word "versus," or "against," is used to separate the names of the parties to the case. In the case of several plaintiffs and/or several defendants, the plaintiffs' names are grouped together preceding the word "versus" and the defendants' names are grouped together following the word "versus."

In the case of a crime, the title of the case would follow the form of "The People v. John Doe."

If a crime is of a serious nature, it is known as a felony. If it is of a minor nature, such as a traffic violation, it is known as a misdemeanor. Whether a given crime is a felony or misdemeanor frequently is set out in the statute defining the crime. Punishment for a felony implies imprisonment and may or may not provide for the payment of a fine. A misdemeanor, on the other hand, will usually require confinement in a "common jail," may or may not require payment of a fine, or payment of a fine may be made in lieu of a jail sentence.

With specific application to the construction industry, Jacob Feld points out in *Lessons from Failures of Concrete Structures*, Monograph No. 1 of the American Concrete Institute, on page 4:

> "Where loss of life is involved in a structural failure, a coroner's investigation is undertaken, and the testimony of expert witnesses is sought, in an effort to establish whether criminal action should be instituted. Financially interested parties frequently organize their own investigations of failures to establish evidence on which to base civil suits."

When a private right has been violated, a civil wrong has been committed. Such a wrong concerns individuals only. The plaintiff in such a lawsuit, the person in whose name the lawsuit is brought, is the individual whose rights have been violated. A suit for a civil wrong might have a title such as "Richard Roe v. John Doe." There are also two types of civil wrongs, breaches of contract and torts.

When two persons have bound each other to certain obligations under a legal agreement, and one of them refuses to live up to the terms of the agreement, a wrong is committed that is known as a "breach of contract." As shall be brought out later, there are several forms of agreement that are considered contractual obligations under the law. A written contract is only one of such forms.

A "tort" constitutes a violation of what are considered an individual's "natural" rights. Such "natural" rights exist in connection with an individual's personal liberty, personal security, and property. Some examples of torts would be false arrest, negligent handling of equipment, or refusal to return a borrowed article.

The usual remedy requested for a civil wrong is an award of money. Sometimes, if a piece of land is involved, there may be a request that title to the property shall be placed by the court in the name of the one adjudged to be the legal owner. In most cases, however, the award is the amount of money that might be required to repair any damages done to the injured individual.

In the light of the previous discussion of the classification of wrongs, here are the pertinent facts concerning the collapse of a portion of the Eastman Kodak Building in Rochester, New York, which occurred on November 21, 1906. It is of interest to note that this is an incident that involves both criminal and civil wrongs.

The design of this building had been changed from time to time during construction, allegedly to protect certain "secret" manufacturing processes. It was one of six similar structures that the same designer-contractor was building. In principle, the structure was a two-story building with brick veneer walls and reinforced concrete columns, slab and girder framing, 134×169 ft. The roof slab was of hollow tile with 4-in. reinforced beams between tiles.

The collapse occurred as the workmen were removing forms and shoring from a section of the northeast corner of the roof. The concrete of the roof slab was about three weeks old and had been pronounced ready for form removal by the construction foremen the day before the collapse. The roof and supporting columns fell without warning over an area 53×88 ft. The mezzanine floor below gave way over a 45×21 ft area. However, the main floor withstood the shock of the falling debris except for a 2-ft square hole punched by a falling girder. Four lives were lost and 15 men were injured.

The owner contended that early form removal was to blame. This contention was based on findings made by his own investigators. There were other inquiries made into the failure. These disclosed a number of faults in the construction work, among these being the following:

1. There was a poor quality of concrete.
2. There were voids and exposed steel in girders.
3. Chips, leaves, sawdust, and blocks of wood were present in the concrete of the columns.
4. The first column to fail below the mezzanine floor was of crumbly concrete, smaller than the designed size and had four reinforcing bars missing.
5. A nearby column had four extra reinforcing bars.

On the basis of the above facts discovered during the various investigations, a coroner's jury brought in a verdict of criminal negligence against the contractor and the owner's assistant manager of works.

HOW DID THE BODY OF THE LAW GROW TO PROVIDE REMEDIES?

The law of the United States came from the "common law" of England. Early English law was unwritten and based on custom. Judges traveled from district to district. In such circuits the court held session for two or three days in each district. During these sessions all pending cases were brought before the judges to be decided. Local customs were recognized by the king's judges in deciding the cases. These decisions and the customs on which the judges based their decisions were recorded for reference in deciding cases having similar facts. From such decisions grew a body of judicial opinions containing expressions of the law. This body of judicial opinions is called the "unwritten law." The customs and usage upon which the judges based their opinions were those which were followed by all the people. References made in legal writings to the "common law" pertain to the customs as expressed in judicial decisions.

Once the highest court of the land had ruled upon a certain set of facts, a precedent was established that took on the force of law. Given exactly similar facts, other judges felt impelled, for the sake of consistency, to give a similar decision. To arrive at a different conclusion would result in confusion in the mind of the public as to what the law really was. This practice of judges standing firm on findings made in the past is known as the "doctrine of state decisis." When the opinion of one judge differs from that of a preceding judge, it is necessary to look for the difference in the facts of the case that might be the cause of the varying decisions.

The fundamental idea behind the common law was that for every wrong there must be a remedy. In many cases, however, the remedies afforded by custom proved inadequate. Exceptional cases for which there seemed to be no adequate remedy were referred to the king. To arrive at an equitable remedy, the king appointed a chancellor to decide such cases. This system of law came to be known as "chancery" or "equity." Some of the cases brought into equity were actions to prevent or restrain a person from doing an act. The restraint that the court of equity would issue were known as "injunctions." Another type of case that was decided by equity was an action specifically impelling a person to live up to his contract rather than merely asking for money damages. This is known as an action for "specific performance." At present, in the United States, the courts of equity and the courts of law are one and the same, and a civil action can ask either for money damages or equitable relief.

Not only is there a body of unwritten law controlling the relationships between people of the same locality, but there is also a body of unwritten law governing relationships between civilized nations. This also has been established by custom and usage and is known as "international law."

Since a major part of the relationships between citizens of various countries occurred as a result of trade between merchants, a similar body of law developed for them. The "law merchant" is recognized by the courts of various countries and has become a part of the established law. Examples of the law merchant are the rules, regulations, and practices connected with the use of negotiable instruments such as promissory notes and drafts.

The entire body of the law has two different phases. The first is involved with the definition of the rights and duties existing in the relationships of individuals with each other, "substantive law." The second deals with remedies, and the enforcement of rights and duties, "adjective law." Let's consider substantive law first.

For example, suppose a contractor has a verbal agreement with a homeowner to place concrete for him for a total amount of $500. This verbal agreement is never put into, or "reduced to," writing. The contractor places the concrete and the homeowner refuses to pay the $500. Can he be forced to pay?

This question is a part of the phase of the law that provides the definition of rights and duties, known as "substantive" law. Commercial law, or business law, is concerned with such rights and duties as arise in the transaction of business. By definition, therefore, it is a part of the substantive law.

In the situation of the homeowner refusing to pay for the placement of the concrete, the contractor has a claim against him for payment. The phase of the law dealing with remedies and the enforcement of rights and duties is known as "adjective" law. This is the primary concern of those engaged in the practice of law itself. It covers such matters as rules for pleading a case in court and the rules of evidence.

Whether substantive or adjective, law may be either written or unwritten. In the early days of civilization, all law was unwritten. Relationships among individuals solidified into generally accepted practices that became custom. Such customs were enforced first by the patriarch, then by the chief, and subsequently by the head man of the community. In time, customs were written down as edicts, codes, or commandments. These, in turn, became the structure of the law.

As the structure of the law increased in size, rules pertaining to various types of situations tended to become grouped into various areas. Before those areas are discussed, it will be helpful to describe the various types of business entities legally able to enter into a contractual relationship.

CHAPTER 2

BUSINESS ORGANIZATIONS

When considering the various types of contractual arrangements and other interrelationships that must be dealt with in the field of concrete construction, it is important to understand the nature of the business entities that are involved in such arrangements. This will enable you to better assess the impact of these relationships on your own business, and assist you in dealing with others.

There are three common forms of organization that a business enterprise may take: (1) a sole proprietorship, (2) a partnership, and (3) a corporation. Each form has its own characteristics, advantages, and disadvantages from an organizational point of view.

SOLE PROPRIETORSHIP

The most elementary form a business may take is that of a sole proprietorship. This is nothing more than doing business for oneself. It is the ownership of an unincorporated business by one person. Generally no legal formalities are necessary to create an enterprise in this form. A sole proprietor owns all of the assets of the business, is entitled to all of the profits, and stands the risk of all the losses, if any, resulting from its operation. He can easily enter and exit from the business as he pleases.

One important element of an organization's existence is its exposure to liability in the course of its business. Unlimited personal liability is a characteristic of the sole proprietorship. Unless limited by express contractual provision with other persons, the sole proprietor is personally liable on all obligations of his business. That liability extends to all of his personal assets, as well as his business assets. Although liability insurance may somewhat reduce the risk of unlimited liability, the sole proprietor remains primarily responsible for his business obligations.

Two other characteristics that are important to a business are related to changes of ownership. "Continuity of business" refers to the ability of an enterprise to continue operating in the event of the death, incapacity, or withdrawal of its owner. "Transferability of interest" refers to the ease with which the ownership of a business may be transferred to another person.

Of necessity the sole proprietorship ends when the sole proprietor sells out or when he dies or otherwise goes out of business. Although the termination of a sole proprietorship upon the death of the sole proprietor may be a disadvantage, provision may be made in his will to give his executor authority to continue the business after death rather than liquidating it. In this way a going business may be carried on by his heirs. In some cases, the ability to terminate or sell a sole proprietorship easily may be considered an advantage.

The management of a business provides another important aspect of an organization's existence. Although the actual management of a business is often the result of factors other than its form of organization, such form does indicate who is responsible for running the business.

There are no formalities as to the operation or management of a sole proprietorship. The sole proprietor is "the boss." Although he may surrender some of his control over the management by express agreement with someone else, the sole proprietor typically remains singly responsible for all aspects of the business operation.

PARTNERSHIP

The key elements of a partnership are the sharing of profits and co-ownership in a business. A partnership is most frequently defined as "an association of two or more persons to carry on as co-owners a business for profit." It is, by nature, an unincorporated business. Although state statutes may provide the basic requirements for its operation, a partnership is created by agreement between or among the partners. The relations of the partners are primarily governed by that agreement.

The partnership agreement may be informal, but a formal contract spelling out in detail the rights and duties of the partners is preferable. The partners must agree to share in the profits of the business, the losses, and in the assets distributed on dissolution. The percentage of the business that each partner owns and a detailed formula for sharing profits and losses should be set forth in the agreement in situations where the interests of the various partners in the business are not equal.

In addition to the rights, duties, and liabilities of the partners arising from the partnership agreement, a fiduciary relationship exists between the partners. Each partner is bound to act in the utmost good faith in all dealings and transactions that affect the other partner or partners in the partnership business. A partner cannot engage in a business that competes with the business of the partnership without the consent of his partners. If a partner reasonably incurs personal liabilities in the ordinary and proper conduct of the business of the partnership, the partnership must indemnify him.

Each partner is an agent of the partnership for the purpose of carrying on its business. The act of any partner in the ordinary course of the partnership's business will generally bind the partnership. Each partner has the right to sign the name of the partnership in connection with partnership business.

As with the sole proprietorship, unlimited personal liability is a characteristic of the partnership. All partners are jointly liable on all claims arising out of the acts of any one of them in the ordinary course of the business of the partnership, as well as for claims arising from the acts of agents and employees of the partnership. Each partner has unlimited personal liability for all of the obligations of the business irrespective of the size of his interest in the business. In other words, he is liable for all claims against the partnership, not just up to the amount of his proportionate interest. Such liability exists regardless of the amount contributed by him to the firm's capital.

Also like the sole proprietorship, the general rule of partnership law is that the death or withdrawal of a partner, or the expiration of the term provided in the partnership agreement, dissolves the partnership, and the affairs of the partnership are to be wound up unless the partnership agreement provides otherwise. However, the contingencies of death, withdrawal, and retirement can be dealt with in the partnership agreement, so that the business may be continued with the remaining partner or partners. In the normal partnership, a partner has no right to sell or transfer his partnership interest to an outsider. This preserves the right of each partner to choose his own partners. While this may be an advantage, it may also complicate the withdrawal of a partner from the partnership.

With regard to management, each partner shares equally in management and business decisions, with the majority rule governing all ordinary matters connected with the partnership business. The partners, however, may agree to a different allocation of voting control or management, and the partnership agreement can delegate the management of the business to one or more managing partners so that the partnership can attain the advantages of centralized management.

CORPORATION

A corporation is created entirely by statute. It is an artificial entity, created and maintained in accordance with the laws of the state in which it is organized. It may be owned by one or more persons. Because it is a separate entity, it can acquire, hold, and convey property, sue and be sued, and generally acts as a separate body in its own name.

The shareholders derive their rights from the corporate charter and the laws of the state in which the company is incorporated. Although the shareholders are the owners of the corporation, they are not agents for it unless they are also officers or employees. They do not owe a fiduciary duty to the corporation unless they are also officers or directors or controlling shareholders in certain situations that may involve a duty of loyalty to the company.

Perhaps the most prominent reason people choose to do business in the corporate form is to achieve limited liability as shareholders. Since the corporation is a separate entity, the liability of the shareholders is limited to their respective stock investments. Consequently, the shareholders, as owners of a corporation, risk only the funds they invest in the stock of the corporation and are not liable personally for the corporation's obligations. In contrast to the unlimited liability of a sole proprietor or a general partner, the shareholder's personal assets are safe from the corporation's creditors.

In addition to providing limited liability for its stockholders, the corporation is also the most suitable form for effecting changes in ownership. The articles of incorporation of most business corporations provide for perpetual existence. Consequently, the corporation continues without interruption on the death or transfer of shares of its shareholders, or its officers or directors. Upon the death of a shareholder, the corporation itself is not divided among his heirs, but continues in existence with only the ownership of the shares being affected. This perpetual existence is a key element of the corporate form of business organization, and generally makes it easier for a corporation to maintain growth, acquire public identification, and attract employees than the other forms of business organization.

The ease of transferability of all or some of a shareholder's shares, without the need for approval by the other shareholders, is another of the principal advantages of the corporate form. This, of course, assumes that there is a ready market for the corporate shares, which may not be true in the case of a small corporation. In addition, shareholders in a small corporation may limit by agreement among themselves the right of a shareholder to transfer shares to an outsider.

Unlike a sole proprietorship or partnership, the method of management and control of the corporation is set forth in the state statute under which the corporation is organized. As a general rule the business and affairs of a corporation are managed by a board of directors. The shareholders owning a majority of the shares of a business corporation usually control the board of directors, but the direct management of the company is by the directors. The management is by the board of directors, with periodic resort to the shareholders for the election of directors and approval of important acts outside the normal course of the corporation's business.

There is, as a result of this formalized system of management, some continuity of policy, and often a more centralized and streamlined operation through the use of the corporate form of business organization rather than other forms of business organization. Although this is usually an advantage in a large business, it may mean nothing in a small business. Moreover, one of the disadvantages of the corporate form for a small business lies in the formalities of its operation, as well as the need to comply with various state requirements, such as annual reports, reports of issuance of shares, franchise taxes, and reports and taxes to other states where it transacts business.

CHAPTER 3
THE CONSTRUCTION CONTRACT

THE RELATIONSHIPS AMONG THE PARTIES

By definition, a contract is a binding agreement between two or more persons or parties. In a construction contract, there are three persons who have a primary interest in the complete and satisfactory execution of the agreement.

The owner. He has bound himself to pay for satisfactorily completed work.

The architect-engineer. He has bound himself to prepare usable and appropriate plans.

The contractor. He has bound himself to produce a structure satisfying the plans.

Each of these parties, upon entering into the agreement, has set certain limitations upon his performance. The owner has set forth the purpose for the structure and the price. The architect-engineer has prepared the plans with this purpose and price range in mind. In effect, he guarantees the structure's suitability if the construction work is according to his plans and specifications. The contractor agrees to do a quality construction job that conforms to the plans and specifications.

The interests and limitations of the parties to a construction contract would seem to be quite clear-cut. Frequently, however, one wonders if there was a genuine meeting of the minds at the time a contract was entered. This is as true with construction contracts as with others. It is particularly important in concrete because the final product is so permanent.

Quoted below is part of one of the articles contained in a construction contract for a government housing project:

> "In any case of discrepancy, either in the figures, in the drawings, or in the specifications, the matter shall promptly be submitted to the contracting officer, who shall promptly obtain a determination in writing from the commissioner. Any adjustment by the eligible builder without such determination shall be at his own risk and expense."

During the actual construction work, three different situations arose that related to this clause. The United States Court of Claims was called upon to interpret the clause relative to each situation in the light of the following questions:

1. Did a discrepancy exist?
2. If so, did the contractor know or should he have known that a discrepancy existed?
3. Was the discrepancy of such scope as to require the contractor to submit the facts to the contracting officer for clarification?
4. Was the contractor obligated to perform the disputed work as part of the contract without seeking clarification?

The following are the three situations in question.

Situation A

The contractor had extensive experience in the fabrication of housing units. He planned to use standard forms for placing uniform stem walls and foundations for 215 housing units. Several foundations had been placed when the contractor was shown large-scale drawings for each house individually. Varying ground elevations were indicated around the perimeter of each house on these drawings. The contractor was directed, therefore, to stop the use of standard forms and to construct individually tailored stem walls and foundations for each house. The contour drawings that were part of the contract documents were scaled so that 1 in. represented 50 ft. The balanced cut and fill of the dirt on the project would be considered consistent with the use of standard forms. However, the use of tailored foundations required the removal of a large surplus of earth.

The contractor's contention was that the use of tailored formwork constituted a change for which he was entitled to extra compensation because:

1. There was no apparent discrepancy in the contract documents requiring submission of the problem for clarification.
2. General practice in the trade and past experience did not indicate conflict, although certain minor differences did exist between the grading and drainage plans and the structural details.

The government's contention was that the contract requirements called for individually tailored foundations and that the contractor must bear the cost involved because:

1. The contractor must have known of the inconsistencies when he submitted his bid under the contract provisions.
2. The contractor had either failed to understand the meaning of the contour drawings or failed to seek clarification.
3. If the contractor had sought clarification, he would have been told that individual forms were required and could have bid on that expectation.

The Court held for the contractor. The contractor's interpretation of the contract documents was within the "zone of reasonableness." The contractor's experience with similar contracts may be considered as a basis of the reasonableness of his interpretation. The scaling of the drawings precluded any specific foundation details, so that any inconsistencies were not so gross and visible as to constitute warning to the contractor that he was obliged to submit them to the contracting officer for determination.

Situation B

The contractor also contended that under the circumstances he was required to plaster walls and to construct larger roof overhangs than the specifications of the contract provided. This entitled him to extra compensation. A reading of the contract provisions and specifications did not correlate in these instances.

The Court held for the government. The direct conflict between the contract provisions and specifications required that the contractor submit the question to the contracting officer for determination. ". . . . plaintiff cannot rely on the principle that ambiguities in contracts written by the government are construed against the drafter."

Situation C

The contract drawings indicated that suitable access to the project existed. During the initial phases of the construction three principle access roads were open and usable. When the government learned, however, that the contractor's laborers intended to place pickets at the access road gates, it directed and limited the contractor's access to the project to a fourth and unsatisfactory road to allow other contractors working on the site free access. Traveling distance and cost of performance of the contract was increased.

The contractor contended that the government had committed a breach of warranty contained in a letter written by the contracting officer that there would be access to the site and the contract provisions that there would be suitable access to the site.*

The Court held that the government had warranted suitable access to the site. That access by one out-of-the-way and unsatisfactory road did constitute breach of warranty. The contractor was entitled, therefore, to damages incurred by being limited to one specific road.

It can be seen that even with careful wording of contract provisions and specifications, as well as a procedure of settling differences of opinion, there can be points in a construction contract on which controversy can arise.

PROTECTIVE WORDING INCORPORATED IN THE SPECIFICATIONS

Most specifications that are part of a construction contract tend to be very precise. There is, however, some tendency on the part of the architect or engineer to protect himself by the use of such terms as "to the satisfaction of the engineer," "in a workman-like manner," or "the contractor shall thoroughly investigate." The inclusion of this type of wording would make any unexpected difficulty the responsibility of the contractor. The contractor presenting a bid on a contract proposal incorporating such

* D & L Construction Co. & Associates v. United States, U.S. Court of Claims, No. 159-65 (November 15, 1968) as cited in *Engineering News-Record*, April 10, 1969.

specifications is faced with the decision as to whether he should provide himself some margin for unexpected difficulties or gamble in his bid that everything will run smoothly.

The requirement that it is the duty of the contractor to check plans can be viewed as an attempt to transfer liability to the contractor for any defects and to obligate him to make corrections. However, there are a series of case decisions that hold that the failure of the contractor to check the sufficiency of plans prior to putting them to use does not excuse the owner from the consequences of an error in the plans.

THE REASONABLY PRUDENT MAN

The fundamental basis of such decisions is the common law rule of the reasonably prudent man. This creature of legal fiction is possessed of the exact degree of skill and wisdom necessary to anticipate all foreseeable circumstances that could occur and to satisfactorily and proficiently take such measures to prevent danger either to himself or others. In measuring the architect's or engineer's liability against that of the reasonably prudent, the engineer must possess such ordinary and reasonable skill usually exercised by one in that profession. Interpretation of this statement by the courts now has approached the point where such ordinary and reasonable skill is, in effect, implied to be a warranty. In other words, the architect has warranted to the owner that the plans and specifications are adequate for the purpose that they are intended to serve. Reliance by the owner on the architect's warranty will shield him from the results of an error on the part of the architect. It will, however, not protect the owner from liability to the contractor unless the contractor himself has not conformed to the rule of the reasonably prudent man.

The owner is not excused from the consequences of an error in the plans unless the plans and specifications furnished are so apparently and obviously defective as to put a contractor of average skill and ordinary prudence on notice that the work cannot be constructed as specified.

We can say that the owner has implied a warranty that the contractor can rely on the plans. This warranty is qualified only by what would be apparent to the reasonably prudent contractor upon visual inspection of the site and the plans.

At first one might say that the wording "in a workman-like manner" would be descriptive of the amount of reasonable prudence a contractor might be expected to exercise. This terminology does not impose upon the contractor any duty for things that ought to have been done, beyond those provided for in the plans and specifications. The application of a principle of law to such a question must follow the determination of the question of fact. This is contrary to the determination of items embodied within the terms of the contract. In such an instance, the principles of law are applied to the terms of the contract for determination of the matter at hand.

Controversies can arise both as to proper execution of the contract and the contract price.

CONTROVERSIES WITH REGARD TO THE SUBJECT MATTER

The majority of the problems involving delay and changes originate at the job site. The initial controversy is usually between the inspector and the contractor's superintendent.

Under the common law, there are limitations placed upon the extent of the inspector's control over a contractor. The contractor's internal operation with regard to his subcontractors, employees, equipment, or method should not be subjected to interference. This limitation, of course, can be altered by the contract itself. The contractor's agreement to a contract containing alterations of the right of noninterference constitutes a waiver that will be upheld by the courts. Such alterations must be set forth precisely or be necessary to produce the results desired under the contract.

Ordinarily specifications state that determination of the intent of the specifications in disputes between the owner and the contractor is the responsibility of the architect. This means that the architect may interpret a specification provision that was not set out clearly. He can, in so interpreting the provision, indicate what he intended when he wrote it. However, he cannot base his interpretation on what he was thinking at the time of the writing alone. The law recognizes only overt acts and does not take thoughts into consideration. His interpretation of the meaning of the words as written can be the only determining factor. In general, the courts will allow such interpretation of only specific parts, not the whole contract.

Where there is an inconsistency between the specifications and intent, the contractor should support his claim by substantial evidence. It must be remembered that the courts have held that without a specific grant by the contract, an inspector has no authority to make changes in the contract or specifications. Without the timely receipt of a written order from the proper authority, the very least that the contractor should do is to submit a protest in writing to the engineer or owner. This protest should set forth that it is his opinion that certain requirements to which he is now expected to comply are beyond the scope of the specifications. The usual reason for an engineer or contracting officer to refuse to furnish a written order is the contention that the required work is within the scope of the specifications. Work embodied in the specifications is work to which the contractor has already given assent. Therefore, it does not constitute additional work or any change in plans. The courts consistently hold that timely notice of a change is required. Unless the contractor can produce written notice or a written order of the proper authority, he would not have a valid case for compensation for additional work. It is the wisest course of action when a contractor is at variance with the engineer or architect on the interpretation of requirements that he await written instructions. His only alternative is to give timely notice of protest. When changes do occur, complete records of time and cost should be maintained. As contracts generally provide that no changes or extra work will be allowed for the purposes of compensation, unless ordered, the burden of proof rests on the contractor.

Disputes arising from the contractual relationships existing between the contractor and his suppliers can also be avoided by means of written orders. The supplier is freed of liability if he can prove that the material supplied was as ordered. It is important to remember that the supplier is not required to verify the specifications unless the order he has received from the contractor requires verification. For example, an order that states that the material is to be "in accordance with the specifications" for the project places the supplier under an obligation or duty to comply in every respect. It is true that the contractor is still primarily responsible under the contract. However, failure of compliance by the supplier may result in liability to the contractor not only for the value of the material itself, but also for the added cost incurred in correction.

Although inspection at the supplier's plant does constitute presumptive proof, it is not conclusive. This means that it possibly, although not probably, could be refuted by other circumstances surrounding the case. If the supplier is obligated to follow the specifications, inspection at his plant is of benefit to him in the avoidance of possible liability. If, as in the case of a ready mix producer, a deviation from the specifications might occur during the shipment from plant to job site, the supplier should keep a record of any variations that do occur. For example, each driver working for the ready mix producer should note the amount of water added at the job site.

There are three basic grounds on which extra compensation can be granted to a contractor without a written order. These grounds can be used as the basis for an award to the contractor even when the contract has stipulated that a written order is necessary. The contractor can use these reasons for his contention that additional money is owed him.

1. The fact that the contract provision may be construed so as to be inapplicable to the particular work for which compensation is sought.
2. The requirement for written notice has been satisfied by some writing other than a formal prior authorization.
3. The fact that the owner has waived or is stopped by his conduct from relying on this formal requirement.

It is important, however, to realize that the burden of proof is always upon the party seeking relief.

As a general statement, we can say that in order for the rights of the parties to a contract to be preserved, procedures should be followed to prevent waiver of rights through inadvertent conduct. All facts required to support a claim should be backed up by adequate records.

CONTROVERSIES REGARDING CONTRACT PRICE

A usual practice is to include a schedule regarding the quantities of work to be done in the invitation to bid. Thereafter, a clause that such quantities represent only approximations and are subject to adjustment is written into the contract.

One aim of a unit price contract is to relieve the burden that can occur through the inaccuracy of the designer's estimate. Valid complaints can arise from major adjustments in particular quantities. Many owners, as well as the federal agencies of both the United States and Canada, generally use the figure 15%. Some states and provincial agencies use higher percentages. The figure that most contractors advance is 10% of the estimated quantity.

If the actual quantity required for completion of the contract exceeds the original contract estimate by more than 15%, the unit price for that item may be renegotiated. Renegotiation must require evidence from the contractor that the increase resulted in an increase in his cost of operation. The presentation of such proof will normally entitle him to an increase in the unit price for that item.

In these days of ever-rising prices, however, it is very possible to lose sight of the fact that while quantities required and costs of operation may increase, there can also be an opposite effect. Circumstances may arise where quantities, and similarly costs, can decrease. In some contracts, it is therefore provided that the owner may obtain an adjustment upon proof that the contractor saved money as a result of changes.

There are several ways in which costs can increase. Normally, if taxes, costs of materials, or certain other factors of costs have risen during the construction period, the contractor will assume this additional cost on the bid quantity. However, the "normal" costs of an overrun quantity might lower the contractor's expected profits to such an extent that he should request some relief. An overrun quantity could necessitate the use of additional equipment to keep up with the contract timetable.

A decrease in quantity could cause an increase in costs, also. The contractor may have transported expensive equipment to the site for work that fails to materialize. In many instances, it is more expensive to process small quantities of materials than to process larger quantities. In such instances, it can be understood that valid complaints can arise as to whom the responsibility for increased costs belongs.

Are there some steps that can be taken to prevent such disputes from arising? As with many other remedial measures employed to adjust differences of opinion, there is no single technique that can be used to avoid all misunderstandings.

Consider the controversy that exists among specification writers as to whether a "method" specification or an "end result" specification might serve to clear the air. With the use of a method specification, a number of the problems associated with unforeseen difficulties increasing the quantity of work required for project completion would be avoided. If the contractor has followed the method specified, he cannot be held responsible for the fact that the timetable was not adhered to or that the quantity of work did not conform to the contract. He, therefore, has a valid claim for any added expense he has been put to in following the prescribed method. However, he is prevented from using any other time- or labor-saving methods to overcome increased difficulties. On the other hand, "end result" specifications would place the entire responsibility upon the contractor's shoulders.

CHAPTER 4

CONTRACTS

For any agreement to produce obligations on the parts of the people agreeing to it, there must be a definite subject matter. It is about this subject matter that the parties will give their assent. They are made aware of their agreement through the subject matter of the contract and its terms.

A precise meaning need not be spelled out in words in some instances. Agreement to one set of terms or conditions can imply agreement to other terms or conditions. The conduct of both parties will imply that an agreement has been reached that has not been reduced to writing. In all cases, a valid contract consists of a definite agreement.

There are procedures available to the parties of the contract to enforce their rights. Similarly, a person entering into the contract can be compelled to fulfill his obligation. In our society, the procedures that are used to enforce the performance of contract terms are those of the law.

In law, the ability to understand the implications of your actions is called competency. The parties to a valid contract must be competent.

A legally enforceable contract is evidence of the rights and duties belonging to one person with respect to others. A company or corporation can be party to a contract as well as an individual.

For the law to take effect, a legally enforceable contract must be valid.

A valid contract is (1) based on legal consideration, or (2) put in writing under seal.

A contract must provide for the performance of some action on the part of the contracting parties. Such an action may be positive, "to do," or negative, "not to do."

Written contract should supersede any and all prior discussions, agreements, offers, etc., between the parties and contain *all* terms and conditions of the contractual relationship between the parties. If one party expects to be paid more for performing obligations under the contract than set forth in its provisions, he should get a *written* agreement to amend the original contract *before* doing any additional work. The contract should contain a provision that it embodies the entire agreement of the parties and can be amended only by a written amendment signed by each of the original parties.

> *A contract is a definite agreement enforceable at law between two or more competent parties based upon a legal consideration or in writing under seal to perform or refrain from performing some legal act.*

A valid, legally enforceable contract is based on:

1. The mutual assent
2. Of competent parties
3. For consideration
4. Regarding legal subject matter
5. In a form legally enforceable for the circumstances.

MUTUAL ASSENT

Mutual assent takes place in two well-defined steps. (1) An offer must be made by one party and (2) accepted by another.

To illustrate this, consider the procedure involved in competitive bidding for a large-scale construction project.

THE OFFER

The requirements of a legal offer are that it (1) must be intended to create a legal obligation, (2) must be certain and definite, and (3) must be communicated to the party who is to accept. The offer may contain stipulations of time, place, or method and may be withdrawn at any time before acceptance.

An offer must be intended to create a legal obligation; thus a promise to attend a dance, or an offer made in jest will not be considered a valid offer by law. However, it must be pointed out that an offer made in jest can have serious consequences leading to legal liability if the person to whom the offer has been made is not aware of the joke.

THE INVITATION TO DO BUSINESS OR TO TRADE

An invitation to do business cannot be considered to be an offer. It is, in fact, preliminary to an offer. An advertisement, circular, or price list is, in effect, an invitation to make an offer.

A contract to be awarded for the construction of a public works project is required by law to be advertised in a newspaper of the locality concerned.

The description of the project itself is usually restricted to a general outline of the principal construction items, and the public is advised as to how and where to obtain such additional information. It is the prepared bid itself that is the offer. Here is the possible wording of such an advertisement.

> Sealed proposals are desired for the performing of all the work and the furnishing of all labor and material incidental to the construction of a community center to be located in City Park, City, County, State, in accordance with the plans and specifications available in the office of the City Engineer.

In some instances, the advertisement itself may carry somewhat more detail for guidance of the prospective bidder. Walter C. Sadler, in *The Specifications and Law on Engineering Works*, gives the following example of a more detailed advertisement:

> "Sealed proposals are desired for the constructing of the intake structures, intake shafts, and river connecting tunnel, at Berry Point, Wisconsin, five miles north of the city limits of Milwaukee. The intake structures will be built of reinforced concrete. The two intake shafts will have an internal diameter of 11 feet and 15 feet, respectively, and will be lined with concrete, the portions passing through earth being of reinforced concrete. The river connecting tunnel will be built in earth at a depth of 100 feet below the surface of the lake. It will be 1,000 feet long, with an internal diameter of 10 feet 3 inches, and it will be lined with concrete."

Some advertisements list the estimated quantities of the principal materials. In the case of reinforced concrete construction, two of the items that might be listed in this way would be cubic yards of concrete and tons of reinforcing steel. The advantage in listing such quantities is to eliminate contractors who would be unable to handle the project.

Another item that might appear in an advertisement is that a performance bond must be submitted with each proposal. Such a requirement may be stated simply, or the requirement may be set forth in detail as:

> Each proposal shall be accompanied by a certified check of 5% of the amount of the proposal, drawn payable to City Board of Park Commissioners of City, to be forfeited to the Board as liquidated damages in case of failure on the part of the successful bidder to execute the contract and furnish satisfactory surety bonds within ten days after the acceptance of the proposal.

Either form of this requirement is acceptable in general practice. Let us take a moment to consider the characteristics of a general offer to do business. It is made to the public in general, but in reality it is addressed to the first person who creates an agreement. Let us assume that in answer to an advertisement, one of the following situations develops:

1. All proposals submitted are too high for the amount appropriated from the public funds for the project.
2. An inexperienced or irresponsible contractor submits the lowest bid.
3. A cheap or unacceptable alternate proposal is submitted.

Any repercussions arising from situations such as these can be avoided by the insertion of a clause regarding the right of rejection. Such a clause might read, "The right is reserved to the Board to accept any proposal, to reject any or all proposals, and to waive defects in proposals."

Detailed engineering plans may be released on payment of a deposit, some or all of which is refunded when the plans are returned in good condition.

Some localities require an affidavit that such proposals are made in good faith and legal intent. Other localities make this requirement a part of their instructions to bidders. Only a few localities require a Certificate of Sincerity to accompany a proposal. When a certificate is required it must be worded so that the proposal is genuine, that it is made in the interest of the bidder, and that the bidder has done nothing to gain an unfair advantage.

In some advertisements a deadline for completion of the work is announced. This is particularly true of large projects where a number of contractors are engaged.

The time of receiving and opening proposals are usually clearly, accurately, and specifically set out in the advertisement. This may either appear at the beginning or end of the wording of the advertisement. It may appear as:

> Sealed proposals will be received by the City Board of Park Commissioners, Room 100, City Hall, City, State, up to 2:00 P.M., Central Standard Time, on the 15th day of March, 19XX, at which time and place all proposals will be publicly opened and compared.

In submitting a proposal, the contractor has the entire responsibility for acquiring the knowledge and information that would affect the terms of his bid. This would include knowledge of climatic conditions, the topography of the site, traffic, access, areas available for storage facilities, and the labor market. Some advertisements require contractors to visit the site of the proposed project. Among the items that may be required are an experience record and a list of available equipment. This requirement is for the benefit of the agency asking for bids in assuring that a contractor has had experience in construction work of the kind required, is fully prepared to perform the contract, and has sufficient financial backing to begin the work promptly and finish in the required time.

Some advertisements for competitive bidding do go into the matter of payment at some length. More general usage is to include this information in the set of general project specifications.

Advertisements for competitive bidding always end with the name and address of the official who is in charge of proposals and letting of the contract. In this way, any additional correspondence or telephone conversations can be specifically directed to the proper party.

It must be stressed once again. *An invitation to do business is not an offer.* In the case of an announcement for competitive bidding, *the bid submitted by the contractor constitutes the offer.*

THE NOTICE TO BIDDERS

A second type of document leading up to the making of an offer is the Notice to Bidders. This has the same status as the advertisement. It is not an offer.

Publicity and talk in general can lead to a great deal of misinformation.

It has been found desirable, therefore, to give a printed list of instructions to the prospective bidders. The instructions carry the title "Notice to Bidders."

The Notice to Bidders includes all the material contained in the advertisement and further elaborates on some of the required details.

One of the principal requirements in the interests of secrecy is that bids be received in a sealed envelope. In order that such a sealed bid not be opened by mistake, the instructions often include the designation that must appear on the outside of the envelope or provide a special envelope to be used. There will also be information as to whether bids are to be submitted by mail or hand delivered.

The instructions will also go into further detail concerning the performance bond. Among the details included are instructions as to how such a check should be made out. Other details will deal with forfeiture provisions as well as the return of such performance bonds to unsuccessful bidders.

The project specifications usually are specific as to payment bonds and performance bonds. It is usually not necessary for the Notice to Bidders to go into lengthy detail regarding such bonds. The Notice does, however, supplement the specifications in setting forth the amount of time allowed for obtaining the required bonds. The Notice generally will also set forth a "save harmless" clause in which the contractor agrees to accept responsibility for expenses, damages, injuries, and loss caused by wrongdoing, misconduct, lack of care or skill, negligence or default upon the part of the contractor, his agents, or employees.

In contract law, an offer can be withdrawn at any time prior to acceptance by the bidder. The competitive bidding procedure is, however, quite expensive, both to the agency and to the contractors submitting proposals. It would be unfortunate if proposals, once received by the agency, could be withdrawn at any time up to the actual signing of the contract. Therefore, the fact that proposals are considered to be irrevocable once submitted, although contrary to contract law, is included in the Notice.

Many private corporations and some government agencies have a list of preferred contractors, and bidding privileges are extended only to those concerns that appear on the list. There are established procedures to be followed for qualification, and once qualified, a contractor is not permitted to submit bids for work of a larger scope. The classes of qualification may be on the order of \$500,000, \$750,000, \$1,000,000. Where such preferred lists are not used, the Notice will generally require that the contractor submit an experience record. A requirement that the contractor indicate the amount of equipment he has available may also be included.

The Notice to Bidders usually will provide a standard form that the contractor can fill in with the price and quantities. The total of these items is the bid price for the job. The Notice will indicate that a blank form is provided, and any proposal not following the instructions may be rejected. It is understood that the estimate of quantities may be approximate, and a statement is made that provision will be included in the contract for increases or decreases in prices. Personal observation of the work conditions and affirming the accuracy of engineering plans (intended to preclude any claims by the contractor of lack of knowledge) may also be required. The legal status of the bidder, as well as qualifications for doing business and the local address of the people having authority to execute a binding contract are to be given.

A detailed time schedule, where applicable, will be set out in the Notice. This is particularly important where completion of one section of the project will enable the next contractor to start.

The Notice invites the contractor to submit alternative methods of construction. An experienced contractor may be aware of a more economic method, and it is to the advantage of the owner or agency to utilize the contractor's skill. Such an invitation does not usually extend to routine work.

The instruction that the bidder inform himself fully is usually expressed in general terms. Detailed instruction can lead to the inference in a court of law that an inadvertent omission is intentional. An irresponsible contractor may try using such a loophole to be excused from responsibility. Such general provisions set forth a method of reporting inconsistencies to the responsible engineer.

The contractor wishing to submit a proposal is at this point informed of the subject matter with which his offer is to be involved. With such knowledge at his disposal, he will then prepare his offer.

An offer may be communicated to the offeree orally, in writing, or by conduct. However, the sealed proposal can only be communicated in the form specified by the Notice to Bidders. As an example of conduct, suppose that a precast concrete producer has a line of garden benches displayed at the entry to his factory. A price is affixed to each model. A customer comes in and hands the owner the amount shown on one of the models and states he wishes to buy that particular bench. The tender of the money constitutes an offer, even though the customer has not stated,

"I will make you an offer of $20 for Model X." If one signs a written offer, he is presumed to have read what he has signed. On the other hand, if trickery has been used in obtaining the signature or terms are made so inconspicuous that the average person would not take notice of them, the signer will not be bound.

An offer once made may be accepted, revoked, or withdrawn at any time prior to acceptance or may lapse because of the expiration of a stipulated time.

Acceptance of the offer can be made only by the person to whom the offer was made (or his authorized agent) and without any qualifications or added conditions. If there are, it is not an acceptance, but in effect a counteroffer that must, in turn, be accepted by the person to whom it is offered. In this instance, the counteroffer is being made to the person making the original offer. An acceptance is legally binding when communicated to the person making the counteroffer.

Let us assume that among all of the proposals submitted to the invitation for competitive bidding, one satisfactory bid was received. The contract will be reduced to writing and submitted to the contractor for signature. What will be the general format of that contract?

THE PARTS OF THE CONTRACT AGREEMENT

The contract agreement is a document, signed by both parties, binding them to a specific obligation.

The title of the document sets forth the type of obligations one would expect to find in the wording that follows. Some contracts set forth the type of duties required such as "Franchise Agreement" or "Partnership Agreement." Others might specifically state the names of the parties involved such as "Partnership Agreement between J. Smith and K. Jones, d.b.a. as Smith-Jones Construction Co." The date of the execution of the contract is included in the title as well.

The introductory clause of the contract clearly identifies the parties, their legal status and the date of execution of the contract. It also makes a statement of fact that the parties have reached the agreement that follows.

The work clause sets out the nature of the duties, or the work, that is to be performed. A great deal of lengthy detail is avoided in the contract by making reference to attached documents that are made a part of the contract, for example, to the specifications and drawings for a construction project.

Miscellaneous provisions are a repetition for emphasis of matters already a part of the contract. Miscellaneous provisions, if they exist, will vary from contract to contract. Unless there is some reason to believe that a matter has not received proper clarification in the attached contract documents, it is best to eliminate miscellaneous clauses.

The duties of the one party have been set forth in the work clause and the obligations of the reimbursing party are set forth in the clauses covering method of payment. Such clauses contain a restatement of the amounts set forth in the accepted offer. Whether there are to be periodic payments or payments made on completion of the various phases of the work are stated in the clauses immediately following those regarding the contract sum.

Within the specifications will be an enumeration of all the documents that constitute the contract papers. Opinion is divided as to whether or not there should be a repetition of this enumeration in the contract itself. Those who feel that some description of the contract documents should be a part of the format of the contract will often include a statement such as, "The advertisement, the notice to bidders, the proposal, the general specifications, the detailed specifications, and the drawings, all of which are attached hereto, together with this agreement, shall form the contract."

All parties to the contract sign the contract. Most contracts are effective without any signatures of witnesses. It is important, however, that any and all signatures applied to the contract show the date of signature.

With the necessary signatures, the contract has become effective, and all parties to it can proceed with what is required of them under its terms.

It is after the contract is signed and work begun that any lack of a genuine meeting of the minds becomes apparent.

TURNKEY CONSTRUCTION

Although neither the term nor the idea behind turnkey construction is new, its use has been expanding during the past few years. Turnkey construction arose from the desire of the general contractor to provide his clients with completed building projects with closely controlled cost and time. Under the turnkey method, designers and builders work together in the development of a design. Preliminary estimates should turn out to be very close to the final costs. In this way, the owner has a realistic cost estimate, and unpleasant surprises become a rarity.

This type of construction represents the ultimate in team approach. All the skills of a seasoned contractor are brought into play. The outstanding feature of the method is that it involves single responsibility for every project. The advantages of competitive bidding are also gained since the greater portion of the construction work is subcontracted on competitive bids.

Contracts will vary, particularly with respect to the way the contractor is compensated. Committees are presently working on the standardization of a basic contract to cover turnkey construction. One of the questions under consideration is fee structure.

After construction has started, the architectural group provides the monitoring and inspection. Inspection procedures are more important than with other types of contracts since the contract manager has full responsibility. The importance of inspection can be seen from the fact that former contract managers become quality control engineers.

There are some pitfalls involved. One is that a general building contractor may attempt design work without fully qualified personnel. This can be disastrous in view of the extent of the legal liability that he has accepted.

The pressure on all members of the organization is intense. Personnel problems, financing, additional overhead, and the fee structure can all present problems. Insurance to cover errors and omissions on such a project is still in the formative stages.

CHAPTER 5

DISCHARGE OF CONTRACTS

The term "discharge" indicates that the obligations underlying a contract have been terminated. The following methods may result in the discharge of a contractual obligation:

1. Performance
2. Mutual agreement
3. Impossibility of performance
4. Operation of law
5. Breach of contract

In event of discharge and termination of contract the rights of parties should be spelled out in contract. They should cover the events that would constitute breach of contract, or would result in automatic termination of the agreement.

PERFORMANCE

Performance is the usual and most satisfactory way for a contract to be discharged. Of the four elements involved in the performance of a contract, time of performance, satisfactory performance, performance by payment, and tender of performance, the most pertinent to the field of concrete construction are those of time and satisfactory performance.

Where no time limit has been stated in the contract, the courts will presume that a reasonable time was intended. There are instances where time is said to be the essence of the contract; if the time provision is ignored, the purpose of the contract may be defeated.

For example, if a contractor arranges for reinforcing steel to be delivered on or before the following Monday and the steel is delivered the following Tuesday, could the contractor refuse to accept? Could he claim it was a day late and he was forced to obtain the steel from another source? Suppose the producer claims that time is not the essence of the contract and delivery of the steel constitutes performance of his entire obligation. Would the contractor have to accept the steel?

Unless the contract provides that performance is to be judged by some specific person, generally performance will be deemed satisfactory if it would satisfy an ordinarily reasonable person.

A masonry contractor agreed to construct three concrete block partitions in the owner's factory, to the owner's satisfaction and approval. When the work was completed, the owner expressed disapproval and dissatisfaction and refused to pay. The contractor proved that the work would satisfy an ordinarily reasonable person. Can the contractor recover?

Yes, he can. The work in fact did not involve personal taste or fancy, such as would be the case in a decorative wall mural. Since it would have satisfied an ordinarily reasonable individual, the owner would have to pay.

There may be instances when the performance of the contract is reasonably, even though not technically, complete. Under the old common law, a party who agreed to an undertaking had to perform that task fully and completely or he could recover nothing. However, with the passage of time, this rule has been modified. The existence of a minor defect will not enable someone receiving the benefit to declare a complete forfeiture. If a contract has been substantially performed, the performing party may require payment, subject to a reasonable allowance for the defects.

For example, a contractor agrees to build a house for an owner. The contractor does build the house, but the owner refuses to pay claiming that there are some minor defects in the plastering. The owner must pay for the building, although he might be able to withhold a small amount to cover the patching of the plastering.

There are instances where the doctrine of substantial performance may be joined to the method of payment in determining the facts of a case.

A building contract carried the provision that the owner would make payments as the work progressed and on written approval of the owner's architect. The concrete contractor applied for a payment without the architect's written approval of his work, and the owner refused to honor the request. Evidence at the resulting trial showed that the architect disapproved the work but had given no reasons for his disapproval and refused to explain why he considered the work unsatisfactory. Would the contractor be able to recover for the work he has done? Consider this in light of the doctrine of substantial performance. As far as the contractor is concerned, he honestly believes he has substantially performed his part of the contract. It is true that he does not have written approval, but the obtaining written approval has been made impossible for reasons unknown.

MUTUAL AGREEMENT

A contract can be discharged by mutual agreement that no party has a claim against any other. The consideration given for such an agreement might be the discharge received from the other. The owner agrees that the contractor is discharged from the contract in exchange for the contractor's agreement that the owner is discharged from the contract.

The terms of a contract may include a provision that the contract is terminated with the occurrence of a certain event. For example, many government contracts state they cannot be assigned to any other party without the consent of the agency. If a contractor does this without the consent of the government agency, the contract is at an end. In fact, the government may force the contractor originally receiving the award to perform as stipulated or be subject to criminal liability.

In common law, the occurrence of "an act of God" did not excuse a party from the performance of a contract. It is true, however, that if both parties agree, the occurrence of an act of God such as a flood or tornado can lead to the discharge of a contract. However, if one of the parties does not consent to such an agreement, it is still in force.

This can also be true regarding unanticipated circumstances such as strikes. Contract provisions may carry clauses providing for alteration of the contract terms should certain specified occurrences take place.

IMPOSSIBILITY OF PERFORMANCE

There are two categories of impossibility of performance. The first is where the subject matter of the contract makes performance physically impossible. It must be remembered that what may be impossible today may be possible tomorrow. Years ago a contract wherein one party agreed to fly to the moon in return for payment would have been considered physically impossible. Today, such a contract is within the realm of possibility.

A second type of discharge occurs when the promissor is incapable of performance. A contract could be made with the services of a certain individual in mind, such as a noted expert in a field in which there are no other practitioners of comparable experience and judgment. In the event this individual dies or is unable to pursue his profession, he is incapable of performing and the contract is discharged.

The promissor may also be incapable of the performance of the contract through the destruction of the subject matter of the contract. For example, if a masonry contractor agreed to erect a series of partitions in a community center, and before he was able to start the building burned to the ground, the contract would be discharged through impossibility of performance caused by the destruction of the subject matter of the contract.

One further area in which impossibility of performance can occur is through a change in codes, regulations, or ordinances governing the construction itself. Let us suppose that a contract has been let to build a detached garage. A zoning ordinance prohibiting detached garages is passed before the structure could be started. The contract is discharged, therefore, by a change in the law.

OPERATION OF LAW

The law itself may provide for the discharge of a contract in four other ways:

1. When a suit is brought regarding a breach of contract and a judgment obtained, the parties no longer have any liabilities with regard to obligations under the contract. This is known as "merger" and the differences of opinion existing between the parties are said to be merged into the judgment.
2. When a contract of "higher security" is created to take the place of one with "inferior security," the inferior contract is said to merge into the superior contract.
3. The best way to make changes in a contract is to provide for such changes in the original document. When an alteration is made by one party without the knowledge of the other, the intent must be con-

sidered. If the intent to defraud is present, the party who is without knowledge of the alteration is freed from all liability under the contract. If the alteration is without intent to defraud, a recovery can be made on the original instrument.

4. The provisions of the national bankruptcy law in the United States provide that the discharge of a bankrupt party from his bankruptcy released that party from all liabilities under any contracts made prior to his bankruptcy.

The time limit for breach of contract is governed by what is known as the "statute of limitations." The time period begins on the day on which the injured party is entitled to bring suit for the claim. However, the time period can be postponed if the defendant is not available to the court either by absence from the state, or by being incapacitated. The statute of limitations will not begin until such incapacity is removed. Examples of incapacity are insanity and imprisonment.

BREACH OF CONTRACT

A contract may be breached by one party failing to perform it or disclosing that he does not intend to fulfill his obligation or making his own performance impossible. If the contract is breached without legal justification, the breach is considered one of failure of performance. If the breaching party has disclosed to the other party that he does not intend to carry through his part of the contract, he has renounced the contract; this is known as anticipatory breach. If the breaching party has committed an act that makes his performance of the contract impossible, this will not excuse performance. For many centuries the common law has held that a person cannot take advantage of his own wrongdoing.

ASSUMPTION OF RISK

A contractor submitted a bid and was awarded a contract. Part of the contract was the excavation for the foundations. In the contractor's bid was an item of $2 per cubic yard for extra excavation. As work progressed, it was discovered that the depth of the foundation would have to be more than the plans had indicated. The contractor thereupon discontinued construction on the basis of unforeseen difficulties. Since he had taken out a surety bond as required by the contract, the contracting agency brought suit against the surety. Decision in the lower court was in favor of the contracting agency. The insurance company, in appealing the case, used the defense that unforeseen difficulties made performance of the contract impossible.

The higher court held that the difficulties in question were no defense for nonperformance and that the insurance company was seeking a pretext for avoiding its obligation. The court reasoned that the contracting business is a speculative activity. The contractor himself provided for the fact that there might be a variation between the plans and the actual construction. It ruled that one engaging in a speculative activity, and aware that some variation might exist, must have the resourcefulness and ingenuity to solve unanticipated conditions.*

In this instance, the contractor made the mistake of not providing an adequate proposal to cover all circumstances that he might encounter in the construction. It is, therefore, a unilateral mistake—a one-sided mistake. Ordinarily, a unilateral mistake will not affect the validity of the contract.

In the construction of a cofferdam on the Columbia River for the United States government, a contractor altered the requirements of the contract. In order to protect a particular phase of the work from unexpected high water, he built up the wall of the cofferdam an additional six feet. The Columbia River is known to be unpredictable in its nature. The contractor was an experienced river contractor. The court held that in view of the facts of the case, the contractor with his past experience and knowledge of the site of the construction work should have arranged his plans to prevent such difficulties.† It might be said that in this instance the contractor did use his resourcefulness and ingenuity to solve an unanticipated condition. However, such a solution amounted to an alibi for not conforming to the requirements of the contract and an alteration of the contract terms. Since there was no fraud involved, the government could require the performance of the original instrument.

A contractor agreed to construct a dry dock for the government for a fixed sum. In his excavation, he unearthed an old sewer. The uncovering of the sewer was anticipated and relocation of the sewer was part of the contract. However, it was in such bad condition and so inadequate that its relocation required more work than anticipated. The court held that the contractor not be excused from performance of the contract nor should he be entitled to extra compensation for unforeseen difficulties.§

* Maryland Casualty Company v. Water Commissioners of the City of Dunkirk, 43 F(2nd) 418.

† Day v. United States, 245 U.S. 159.

§ United States v. Spearin, 248 U.S. 132, 39 S. Ct. 59, 63 L. Ed. 166.

It can be generally said that contractors engaged in underground construction assume the risk for uncovering unexpected sewers, gas lines, electrical conduits, or basements extended under the street. Such difficulties should be anticipated.

The courts have found little sympathy for those who wish to avoid or change their contractual obligations because the work became more difficult. The bid price should include the risk assumed since it is the contractor who determines the price at which he expects to do the work.

To what extent will the courts hold that certain risks are inherent in the pursuit of any given activity? Anyone participating in an occupation is assumed to know that certain difficulties may exist. Knowing this he is assumed to be able to take necessary precautions to avoid nonperformance of his contract. This attitude of the court extends beyond the assumption of knowledge regarding difficulties in working conditions.

In instances where weather might affect the performance of a contract, the bidder assumes the risk for adverse weather conditions. In concrete construction this is particularly pertinent. Consider the precautions that must be taken when concreting in hot or cold weather. In actual fact, severity of weather conditions does not consitute a good defense against a suit for nonperformance.

What about labor difficulties? In one instance, the steelworkers on a construction job went on strike for a period of nine weeks. Would this be an instance where knowledge would be assumed and precautions required? The court felt that it was and that the strike was no defense for nonperformance. The contractor should have anticipated all potential sources of labor troubles, and it was up to him to have taken the necessary precautions before submitting his proposal.* Would this reasoning apply if the strike involved workers in another industry, such as the gasoline industry, and not those employed by the contractor himself? It is felt that this might be dependent upon the amount of public knowledge regarding unrest in the striking industry. Wildcat strikes are unanticipated or unexpected. Labor negotiations give rise to the suspicion that a strike condition might exist in the near future. There have been instances where a contract provision has been inserted to avoid such repercussion effects.

The breakdown of machinery and equipment is not cause for nonperformance. The court has held that a contractor has the responsibility to have the men and machines to do the job he contracted. The contractor

* Schaeffer v. Brunswick Laundry, 183 A. 175, 116 N. J.L. 268.

is assumed to have knowledge of what kind of mechanical difficulties can occur and is expected to anticipate them and prevent them.*

Another defense for nonperformance that has proved to be faulty is augmented difficulties. For example, suppose that a contractor entered into a contract for supply of aggregate from a certain source. The supplier of the aggregate finds that removal of aggregate from that source has become burdensome because costs have gone up. Such "augmented difficulties" are no excuse for the supplier to refuse to deliver the aggregate from that source. It is not the function of the court to decide whether the parties to a contract acted wisely or rashly. Contracts are made to be performed, not to be broken, and the law will not be utilized to defeat the objectives of agreements.

We can summarize, therefore, some of the areas in which the contractor undertakes the assumption of risk in submitting a proposal or entering into a contract. These are:

1. The risks involved in a speculative activity
2. Unilateral mistake
3. Past experience and knowledge
4. Risks inherent in one's chosen line of business
   a. Working conditions
   b. Labor conditions
   c. Operation of machinery
   d. Weather conditions
5. Augmented difficulties
6. Poor judgment

RESTRICTIVE CLAUSES

In attempting to minimize some of the effects that might occur through the assumption of these risks some contracts incorporate clauses that will in effect excuse the contractor from performance if certain unanticipated events occur. Two areas in which the contractor may not be possessed of any foreknowledge are the "act of God" and the "vis major." An "act of God" is an event that occurs through an irresistible force of nature such as flood or tornado. A "vis major" is the result of a human unrest such as war or a national strike.

An "act of God" does not excuse nonperformance. What is the court's reasoning in this matter? A contractor had entered into a contract to con-

* Porto Rico Sugar Company v. Lorenzo, 32 S. Ct. 133, 222 U.S. 481, 46 L. Ed. 277.

struct a schoolhouse by the end of May. During the construction the building was struck by lightning and the contractor was prevented from finishing construction by the end of May. A tenant is required to return the land in the same condition as he received it but is not held liable if prevented from doing so by an "act of God." However, although a contractor is performing work on land belonging to the owner and that land is visited by an "act of God," it has been held that until completion of the job and delivery to the owner, all labor and material belong to the contractor. Therefore, the construction contract constituted an unconditional promise and an "act of God" was no excuse for nonperformance. The court concluded that any contractor who does not want to assume the risk of losing his material and labor during a construction job should carry insurance.*

This decision points out one way in which a contractor can seek to minimize his risks.

Let us suppose that a form which a contractor signs ordering supplies contains the following wording printed upon it, "All agreements contingent upon strikes, accidents, and other unavoidable delays, beyond our control." Would the contractor have a cause for action if the supplier failed to deliver because of such an occurrence? In such an instance, the court will carefully investigate the underlying facts to determine whether the delay was indeed beyond the control of the nonperforming party. If the delay was occasioned by the breakdown of dilapidated machinery, it can be assumed that the supplier might have reason to believe that a delay would occur and if so, that delay could have been prevented by appropriate action on the part of the supplier itself. The underlying factors would determine the court's decision.

The following is a case where the underlying factors were so unprecedented as to be unanticipated by the contracting parties when the agreement was made. A contract to construct a levee contained the wording, "All damages or injury to work, resulting from floods or other causes before the work has been received by the contracting officer, shall be sustained by the contractor." A flood of unprecedented proportions occurs. The flooding that had been anticipated at the time of the signing of the contract was expected to wipe out no more than the construction work. However, the entire levee was demolished. The holding of the court was that regardless of the extent of the risk to which the contractor assumed he was agreeing, he had made a definite and specific agreement. The

* School District No. 1 v. Dauchy, 25 Conn. 530, 535 68 A.M.D. 371.

United States was entitled, therefore, to full performance of the original contract without additional reimbursement to the contractor.*

The foregoing situations may seem to indicate that the law can be unduly hard on the contractor. There are, however, instances where such a mistake in a contract turns out to grant a remedy.

CONTRACTUAL MISTAKE

If an agreement would not have been reached if the facts had been known, a contractual mistake may be claimed. There are several remedies for a contractual mistake.

1. The contract may be corrected to remove the error.
2. The contract may be cancelled with such additional relief being granted to restore the rights of the injured party.
3. If no performance has taken place, the contract can be discharged by mutual agreement or court order.

The contract can be corrected only in cases where none of the parties to the contract will be injured by the removal of the error. As everyone is assumed to know the law, a mistake caused by a misconception of the law offers no defense.

However, there are certain types of mutual mistakes where there can be a remedy by law.

One type of mutual mistake concerns the subject matter of the contract. A contractor signs a leasing agreement for a piece of equipment. The salesman executing the agreement has been told that the equipment is immediately available. Unknown to the salesman the equipment has been leased and delivered to another firm. The salesman honestly believed the equipment was available. There is no other equipment on hand at the leasing company that can perform the specialized function. If no money has passed, the contract is void. If money has passed, the contract can be rescinded and the money returned to the contractor.

A contractor, ordering supplies from a supplier, signs a note in blank. Since the prices to be paid for the supplies have already been agreed upon, he believes that the note will be made out in the correct amount. By error the amount typed on the note is several hundred dollars more than the amount the contractor agreed to pay. If the supplier agrees, the court will order correction of the note to the amount agreed on, calling it

* United States v. Lewis, 237 Fed. 80, 150 C.A.A. 282.

an instance of mutual error. If the supplier were to insist upon the incorrect amount, an action could arise carrying with it criminal liability for changing the face value of the note.

It must be pointed out that for a recovery to be made on mutual mistake, the mistake must be concerned with the essence of the contract and not with a side issue. A contract was entered into between the owner and a contractor for the driving of foundation piling. Both parties thought that the underlying material was clay and sand, which would make penetration comparatively easy. However, when the work was started it was discovered that there was a great amount of riprap and heavy debris in the substrata. The contractor sought reimbursement for the additional work involved. The court held that the subject matter of the contract was the driving of the piling for a stated price, and that no place in the contract was any mention made of the substrata material. Furthermore, the court stated that had the driving of the piles been easier than anticipated, the owner would not have gotten a refund from the contractor.*

Generally, the courts will limit relief to mistakes of fact about past or present events, but not for unexpected future developments. One cannot ask that a mutual contractual mistake be declared for what might happen.

Correction of a contract will be granted in the case of mutual mistake. Although the courts will cancel a contract resulting from unilateral mistake, they will not order a correction of the contract for this reason. To do so would not be embodying the intentions of the second party to the contract. A mutual mistake indicates that there has been no fraud or negligence on the part of either party. Since two parties are involved, the burden for the correction of the error is on the one whose mistake was the most contributing cause for the error.

When considering unilateral mistakes, it must be noted that the party bringing the action into court, the plaintiff, must not have caused the mistake through his own negligence or indifference.

A contractor entered into a contract with a city to construct a tunnel. The city had prepared plans of the tunnel and included in such plans a profile of the subsoil indicating that the tunnel would go through solid rock. The specifications required that the contractor make his own tests of the ground before he made his bid. As work progressed, it was found that the subsoil material was composed of clay, boulders, and gravel, which required more expensive methods of construction. The contractor brought suit requesting that the city pay $25,000 over the $45,000 con-

* Cavanagh v. Tyson, Weare, and Marshall, 116 N.E. 818.

tract price. The court held that it was the contractor's own negligence in not performing tests of the subsoil that brought about the increased costs; he therefore was not entitled to additional compensation.*

There are several actions that a party can take when he discovers that he has made a mistake in a contract. If the error is small, it might be best to ignore it. If not, he may (1) submit a claim for extra compensation, but continue to fully perform the contract; (2) attempt to renegotiate the existing contract; (3) if a mutual mistake is involved, he may bring a court action to correct the error, alleging damages if any; or, (4) he may bring an immediate court action for cancellation of the contract, alleging damages, if any. In any event, he must act promptly when discovering that a mistake has been made.

A contractor submitted a proposal for constructing a water main. An error was made in completing the proposal, and tunnel excavation was priced at $.50 per cubic yard instead of $12 as intended. A $90,000 bond was required to be submitted with the proposal. The contractor immediately brought an action into court to obtain relief. The city, ignoring the protests of the contractor, let the contract to the next bidder and maintained that the $90,000 bond was forfeited. Since the contract had been awarded to another, there was no possibility of correcting the agreement signed between the plaintiff and the city. The contractor had indicated good faith by his prompt notification of the error, and the court directed a cancellation of the contract and a return of the bond. Not only did the contractor save himself a monetary loss, but the prompt action also saved the second contractor from being disadvantaged.†

In this instance, the contractor's employee had apparently made an error in preparing the proposal. There are many instances where the actions of an employee can bind his employer to a contractual obligation. There are, of course, many shades of meaning to be attached to such actions and, as in contract law, the circumstances of each case determine the court's decision.

* City of Dallas v. Shortall, 114 S.W. (2d)536.
† Moflett, Hodgkins, and Clark Company v. City of Rochester, 44L. Ed. 1108.

CHAPTER 6
THE LAW OF AGENCY

The law of agency is based upon two fundamental concepts. The first is that one person (the agent) may act for and in the place of another person (the principal). The second is that the acts of the agent are considered the acts of the principal whenever the agent is known to be acting in that capacity. As technology has increased over the years, it has become more apparent that one person cannot be an expert in all areas; for this reason specialists are employed. It is also apparent that one person cannot do everything connected with some of the immense projects being built. He must employ other people to help him with the various tasks. However, both the experts and the workers are, in effect, acting for their employer.

To prevent any dispute over the terms of agency relationship, the scope of an agent's duties and authority should be spelled out *in writing* —if only by a letter.

TYPES OF RELATIONSHIPS

One can readily understand how this applies to a construction project. What one cannot see at first glance is that on a construction project there are, in effect, three relationships relevant to the law of agency. They are:

1. Owner and independent contractor
2. Principal and agent
3. Master and servant

The ability to distinguish these three classes can be important since such a distinction will determine the responsibility and liability for all parties concerned. For example, if there are unpaid bills for materials in the absence of any statute altering the liability, who is liable—the owner or the contractor? Primary liability rests upon the contractor. However, should the contractor default in the payment of these bills, the liability is with the owner. Principal will be liable for a contract executed by his agent. However, there may be circumstances where civil wrongs committed by the agent will not vest liability on the principal. It has long been the law that the master is responsible for the negligence of his servants; the employer is responsible for the failings of his employee.

Certain tests and definitions have been devised to distinguish what type of relationship exists in different situations. However, the complexity of factors that may exist prevent any single rule for determining relative responsibility.

INDEPENDENT CONTRACTORS

The determination of whether a person is acting as an agent or an independent contractor may depend on the amount of control he exercises over how the work is to be accomplished. The independent contractor is responsible for the results and attends to all details necessary for obtaining those results. He establishes the work schedule, chooses the method by which it is to be done, obtains the various materials, and is paid for the end result. In an action where someone is attempting to establish that the defendant acted as an agent for the owner, the owner may use these factors to show that the defendant is in reality an independent contractor and liable for his own wrongdoings. It might be said to be that control over the duties performed establishes an independent contractor status.

Consider the following example. As a result of negligent driving, the owner of a station wagon injures a pedestrian. At the time the accident happened, the driver, who was employed full-time by a contractor, had been sent to the building material supplier to pick up some hand tools. He had been told by the contractor which supplier to go to and the easiest way to get there, and he was to pay for the tools with the contractor's personal check. The station wagon belonged to the driver and the contractor had no personal knowledge of the driver's ability in managing the vehicle. Would the contractor be liable? Yes. The court would say that for the hours involved the driver was obligated to be working at his job, and the contractor had designated his route and task.

AGENTS

By definition, an agent is one appointed by a principal to act for him in some transaction or series of transactions with respect to third parties. The principal, in turn, is the party for whom an agent acts and who will receive the benefits of the agent's actions with respect to third parties.

In the light of what has just been discussed, what would be the relationships in the following situations?

The owner engages an architect to design an enclosed shopping mall (owner-independent contractor). Upon receiving the designs, the owner is enthused and asks the architect to have complete specifications drafted, and to get in touch with the contractors who have proven satisfactory in his (the architect's) past experience and ask them to submit bids. (With regard to the specifications, the architect is an independent contractor. However, in contacting the contractors, he is acting as the owner's agent.)

There are two usual classifications of agents. The general agent is one who has the authority to carry on all of his principal's business of a certain kind or in a given locality. A special agent is one who is employed for a specific purpose or particular act or acts and no more. All persons with whom an agent deals on behalf of his principal are known as third parties. They are outside of the agency relationship itself. In communicating with the contractors, the architect is acting as a special agent. His agency in this respect is terminated when the requests for proposals have been made.

Since the design requires some very sophisticated concrete construction, the owner asks that the architect receive the bids himself. After he has had a chance to go over all the proposals, the owner, being unfamiliar with the ins-and-outs of construction, will continue to follow the architect's advice and retain his services throughout the construction of the shopping mall. The verbal agreement reached between the owner and the architect was reduced to writing. The contract contained assurances that the designs and specifications would be satisfactory as well as all the other provisions that are usual for this type of agreement. A set fee had been established for the design as well as an honorarium for the services of the architect in getting the contractors to bid. The contract included reference to the design and assistance that were in the prior agreements. (A merger of the inferior contracts has been made into the final executed contract.)

An agency may be created by appointment. An appointment consists of the delegation of a specific task by a principal to an agent. The original request of the owner that the architect ask for bids from contractors was an appointment.

The owner received a bill from a structural engineer for professional services rendered prior to the date of the written contract. The wording of the bill indicates that such services were rendered in connection with the design of the shopping mall. In checking with the architect, the owner discovered that the architect had approached an acquaintance who was a structural engineer some time before he had submitted his design to the owner. The gist of the conversation between the architect and engineer was that the architect had stated he was acting as the owner's agent in preparing the design and specifications for a shopping mall. (At the time the architect was engaged by the owner to prepare the plans. A statement that he was acting for the owner was done without any authority from the owner.) His conception of the overall design required the benefit of the engineer's technical knowledge of the strength of structural members. In fact, he had based the design of some of the structural members shown in his drawings on information received from the structural engineer.

The person on whose behalf an unauthorized act is done may (1) repudiate it, (2) expressly ratify it, or (3) ratify it by conduct. To accept the benefits of an unauthorized act with full knowledge of the circumstances is ratification of an unauthorized act, whether such an act is without any authority or merely in excess of the authority officially granted.

The specific rules for the ratification of an unauthorized act may be summarized briefly. The only person who can ratify an unauthorized act is the one in whose name the act has been done. The principal ratifying an unauthorized act must have full knowledge of all the facts and circumstances. The principal can ratify whatever he can lawfully approve. A principal cannot ratify the profitable portion of an unauthorized transaction and repudiate the unprofitable. The ratification must cover the entire act.

Having been made aware of the facts concerning the cooperation between the architect and engineer, the owner ratified the architect's statement that the request for help was made on the owner's behalf by paying the structural engineer for his professional services (ratified by conduct).

Before receiving the original designs from the architect, the owner had toured several public expositions at which the manufacturers of various building materials had exhibits. Noting his interest in several items, one of the manufacturers began a selling campaign. During one of the visits from the salesman, the owner inferred that the salesman should see the architect. Subsequent sales efforts were, therefore, directed to the architect as agent for the builder. In a case involving the manufacturer, the owner would be barred from denying that the architect is his agent because he has inferred that by dealing with the architect the manufacturer is dealing with his agent. Any contract between the architect and manufacturer would bind the owner as principal.

During the progress of the work at the architect's office, the stationery supplies ran out. The architect asked his secretary to go to the office supply store and stock up on everything she thought they would need for the next few weeks. At the stationer's, the secretary ordered the necessary supplies, stated that she worked for the architect, and asked the stationer to bill her employer. To the stationer, the secretary is acting as an agent in binding her employer to a contractual obligation to pay for the supplies. This is superior to the relationship of employer-employee that arose from the old common law relationship of master-servant.

DUTIES OF AGENTS AND PRINCIPALS

Following the execution of the contract between the owner and the architect, and upon the architect's recommendation, the owner entered into a contract with the structural engineer. In this contract it was set out that the engineer would work in conjunction with the architect in drawing up project specifications and providing inspection and approval as construction continued. To fulfill this contract, the engineer organized a staff of skilled technologists including an inspection team. The specifications as finally drawn up included provisions that the engineer's approval of the construction was required and that the inspectors were responsible to the engineer. In this way, the inspection team would be in fact the engineer's agents.

The duties that the engineer owed to the members of the inspection organization as his agents were compensation for services rendered, reimbursement for expenses reasonably incurred, and an indemnity against such risks as the inspectors would reasonably and necessarily assume in the proper discharge of their duties.

On the other hand, the members of the inspection team, as agents of the engineer, owed the engineer the duties of loyalty and good faith, obedience to instructions, the exercise of skill, care, and diligence, the keeping of accurate records and accounting to the engineer for actions taken in the line of duty. The members of the inspection team were advised that only in such cases where it would be absolutely necessary for the proper discharge of duties could one member of the team appoint another person to take over his duties. In cases of absolute necessity, the subagent so appointed would automatically become the engineer's (and in effect, the principal's) agent. If a member of the inspection organization appoints a subagent when such an appointment was not a case of necessity, it would constitute a breach of good faith, care, and diligence.

Let us now consider three actual cases in the light of the brief survey we have made of agency law.

The first situation involves a contract with a state highway department for a riprapping operation. The contractor's proposal on which the contract was awarded fixed the price at $1.84 per cu yd for approximately 12,000 cu yd, using what is termed "one-man" stone. The contractor agreed to furnish all labor, materials, and to conform to standard state specifications. When the state engineer visited the site, he directed the contractor to use "two-man" stone, which is heavier and more costly than that specified in the contract. The contractor brought suit for the extra expense involved. His contention was that, notwithstanding the contract terms, he was justified in relying on the state engineer's directions and the state was bound by its own representative. The state contended that it was not liable as the contract terms did not grant the engineer authority to modify its terms.*

The court held that the state was not liable. Since the additional requirements of the state engineer were not covered by the contract, failure to follow the state engineer's directions would not have placed the contractor in default.

Our second actual case is concerned with a contract to rebuild a pier destroyed by flood. The new pier was to be at the exact location of the former pier. The contract provided that the county engineer, if requested

* Grifis et al. v. State, 138 N.W. 2nd 138, Sup. Ct. of South Dakota as cited in *Contractors and Engineers Magazine,* April 1969.

to do so by the contractor, would place stakes to indicate the location of the new pier. As an alternative, the location could also be determined by agents employed by the contractor. The contractor, using his option, requested the services of the county engineer in placing the stakes.

After a considerable amount of work had been done, it was discovered that the location was in error by 28 ft. Work was stopped and restarted at the correct site only after a long delay.

The contractor brought suit for damages, based on the fact that he had relied on the work of the county engineer. The county contended that it was the contractor's duty to carefully check the site location,* and, therefore, the negligence was his.

The court held for the contractor. Under the provisions of the contract, the contractor had a right to rely on the information furnished by the county engineer.

A principal incurs no obligation toward a third party unless he has authorized or ratified the agent's acts. However, such authority need not be expressly set forth—it can be implied. In the case of a general agent, if an act is within the apparent scope of that agent's responsibility, the principal will be bound by the agent's act. The reasoning behind this is that the principal has placed the agent in a position so that third parties would be unaware of the fact that the agent was committing an unauthorized act. In the case of a special agent, the principal will not be bound by any act of the agent outside the scope of his authority. Third parties, when dealing with a special agent, should investigate the extent of the agent's authority. This is true when the agent states that he does have the authority.

A contractor entered into a written contract for the construction of a motel and restaurant. The contract specified that all change orders be in writing and signed by either the owner or the architect.

The owner's agent made numerous verbal requests to the contractor for changes in the work. None of these requests were put in writing. The contractor informed the owner's agent that additional costs would be incurred. He did not proceed with such changes until he received the agent's promise to pay for the additional work.

The owner refused to pay the extra costs on the grounds that it was clearly specified in the contract that no extras would be allowed unless a signed, written change order existed.

* Board of County Commissioners v. William J. Howard, 230 F. 2nd 561 (U.S. Ct. of Appeals, 10th Cir.) as cited in *Contractors and Engineers Magazine,* April 1969.

The decision of the court was that the contractor was entitled to the additional costs for performing extra and changed work. This determination was made even though the authorization required by the contract for the extra work did not exist.

The logic used by the Pennsylvania Supreme Court* was as follows:

1. The owner's agent requested the changes and promised to make payment.
2. The agent was often at the site and saw the extra work being performed.
3. The agent knew that the extra work was being performed without proper authorization, but did not raise any objection.
4. The agent's conduct constituted an implied promise to pay for the extra work.
5. The extra work was performed, therefore, either under an oral contract, separate and distinct from the written contract, or under an implied agreement that the written authorization requirement of the original contract had been waived.
6. The owner cannot refuse payment on the technicality that the written authorization required by the original contract did not exist.

DISCLOSURE OF PRINCIPAL

Let us return to the transactions between our architect and the owner of the shopping mall. You will recall that the owner asked the architect to contact various building contractors. Let us suppose that the architect did contact several contractors. To some of them he did not mention that the project belonged to the owner, and they had proceeded in the supposition that they were preparing estimates for the architect. To others, he mentioned that the construction was in behalf of the owner and that he was operating as the owner's agent.

In the first instance, the principal is undisclosed. If the third party later discovers the identity of the principal, the third party can choose to hold either the agent or the undisclosed principal liable for any injuries or damages he may be subject to through dealing with the agent. He cannot, however, hold both of them liable. If he elects to hold one liable, the other is free of responsibility.

In the second instance, the principal is disclosed, and the third party, therefore, can only hold the principal liable.

* Universal Builders, Inc. v. Moon Motor Lodge, Inc., 430 Pa. 550, 244 A, 2nd 10 (1968) cited in *Engineering News-Record,* May 22, 1969, page 105.

If the undisclosed principal terminates the agency and settles with his agent before his identity is discovered, he is no longer exposed to liability.

There are some aspects of liability to which both principal and agent may be subjected.

If, in the course of his duties, an agent commits a crime, the principal is as liable as he would be if he had committed the crime himself. This is true also of torts (civil wrongs). If such acts are committed outside the scope of the agent's duties, however, the principal is not liable—only the agent. A principal will be liable for his agent's frauds if they (1) are committed within the course and scope of the agent's duties, (2) are ratified by the principal with full knowledge of the facts, (3) are induced by or participated in by the principal. This is true also of misdemeanors and felonies.

An agent may become liable to third parties under the following circumstances:

1. When he represents that he acts as agent for another, either without authority or in excess of his authority.
2. When he conceals the fact of his agency and deals with a third party as if he were a principal.
3. When he commits a tort, either within or outside the scope of his duties as agent.
4. When he represents an incompetent principal.

Agency may be terminated in two different ways, either by act of the parties or through the operation of law.

As in contract law, the termination of agency by the parties may be through complete performance of the terms of the agency, mutual agreement, revocation, or abandonment.

The termination of an agency through operation of the law will occur when the subject matter of the agency becomes illegal; or when the subject matter of the agency is lost or destroyed; or through the death or incapacity of any of the parties; or through the dissolution or an adjudication of bankruptcy of either a business firm or individual.

The only time that an agency is irrevocable is when the agency is coupled with the interest of the agent in the subject matter of the agency. Such an agency will continue despite the death, incapacity, or bankruptcy of the principal.

CHAPTER 7

EMPLOYER-EMPLOYEE RELATIONSHIPS

Today, the relationship between employer and employee is a contractual one. The contract may be a hiring contract, either express or implied. This description of the relationship has not always been true. The early character of the association existing between employer and employee was designated as master-servant.

NATURE AND HISTORY

In the master-servant relationship, the employer was the master in every sense of the word. His authority was not to be questioned. The employee was in fact a servant completely subject to the will of the master.

In contrast to this, today, whether an employee acts independently or in conjunction with a trade or industrial union, his relationship to his employer is governed by the principles of contract law.

Employment, as such, is a contractual relationship, express or implied, but as we think back, agency is a contractual relationship, express or implied, as well. What is the standard that can be used in characterizing whether a business relationship is one of agency or employment?

Insofar as third parties are concerned, business transactions entered into by an agent are equal to business transactions entered into by the principal. The violations of the rights of third parties by an agent in the course of his duties as agent are violations by the principal. Therefore, the agent can legally bind his principal by his actions.

It is true that the sale of merchandise by a clerk in a store has the effect of binding obligation upon that clerk's principal, the store owner. For instance, the terms on which such a sale is to be made (the selling price, quality, etc.) are not within the clerk's discretion but have been prescribed by his principal. The employee may bind the principal within the scope of his employment, but the scope of that employment is strictly prescribed. Generally speaking, therefore, the distinction that exists between agents and employees is that agents have wider authority and discretion than do employees.

The hiring contract, whether express or implied, must have all of the essential elements of a contract—mutual assent, competent parties, consideration, and valid subject matter. If, at the start of employment, it is the understanding of both parties that the employee is to work for more than a year, the contract must be in writing. Many employees who have worked for many years for a firm have never signed an employment contract. However, it was not a part of the terms of their employment that they promised to work for any specified period of time. Let us consider, however, the case of an engineering consultant who has been employed by a contractor. In the employment negotiations preceding the employment, the parties agreed that the engineer would work for two years as a consultant. The contractor could not dispense with his services before the two years' time without cause and without some satisfactory arrangements being made by mutual agreement on the remainder of the contract. On the other hand, the engineer could not leave the employment of the contractor before two years' time (unless his performance under the contract has been rendered impossible) without being liable to the contractor for damages. Such a contract of employment for a period longer than one year must be in writing.

The hiring contract that usually governs the employee's relationship with his employer is implied. During the course of employment interviews, agreement is reached verbally as to pay, paydays, duties, hours, fringe benefits, etc.

LABOR LAWS

In the earlier relationship of master and servant, the master's authority was not to be questioned. However, the interest of the government in the social welfare of the people in general has led to the enactment of laws concerning labor conditions. The rights and duties of any contract of employment are subject to the provisions of such laws as govern in any given locality.

The more important labor laws concern (1) factory laws and working conditions; (2) wages and hours; (3) collective bargaining; (4) child labor; (5) social security; and (6) workmen's compensation.

Buildings used for factory and business purposes are required to be inspected periodically. The sturdiness of the building is checked by inspecting the foundation, beams, and floors to ascertain that the structure can withstand the loads and stresses to which it is subjected. The possible existence of fire hazards and the adequacy of exits are also given considerable attention. There are rules governing sanitation and ventilation practices to be used in working areas. To prevent mechanical or occupational hazards there are elaborate precautions set forth in various laws regarding safety devices and regulating or prohibiting the manufacture of products that involve harmful or poisonous substances.

Federal legislation was originally enacted regarding wages and hours in connection with the employment of women and children in industry. As early as 1923 Congress enacted minimum wages for women and children. However, this was subsequently declared unconstitutional and state statutes that had been drafted along similar lines were declared void by the courts. In 1937, however, the United States Supreme Court overruled its prior decisions and sustained the constitutionality of a minimum wage law for women and children that had been adopted by the State of Washington. This opened the way for a national statute, and in 1938, the Fair Labor Standards Act, or Wage and Hour Law came into being.

The constitutionality of the Fair Labor Standards Act has been reaffirmed by a Supreme Court decision handed down in 1941. The general provisions of the Act laid down a floor under which hourly wages may not descend and established a ceiling on the number of hours comprising a working week. The use of child labor under certain ages, depending on the type of work, is prohibited. Any exceptions to such prohibitions are specifically set forth. The Act also establishes that the provisions it contains shall be administered by the Wage and Hour Division of the Department of Labor. State statutes pertaining to these matters are modeled after the federal statute.

Many labor laws have arisen out of the fact that during the Industrial Revolution, in the early part of the nineteenth century, children as young as seven worked from 12 to 16 hours a day in the growing factory system.

The evils reflected in the disregard for the health and safety of these children became a matter of public shame. Child labor laws began with compulsory school attendance and the prohibition of the employment of children under 14 during the school term. Factory laws were also enacted that limited the employment of children as to age and hours of work.

In time, the public conscience was directed to a nationwide program of social and economic security. The Social Security Act originally applied to persons employed in factories, shops, mines, mills, stores, offices, banks, building and loan associations, ships of American registry, and others. It has, since 1950, been enlarged to include domestic servants, farm workers, and the self-employed.

Those states cooperating with the federal government under the provisions of the Social Security Act provide assistance to members of the working force of this country in three ways. There is assurance of financial assistance for those eligible in cases of unemployment, old age, and survivorship. The second type is the public assistance of the aged, the needy blind, and aid to dependent children. The third area encompassed within the Social Security Act is that of health and welfare services, including child welfare services, services for crippled children, maternal and child services, and public health.

As the historians of industry have referred to the nineteenth century as the Age of the Industrial Revolution, in the future they will refer to our twentieth century as the beginning of the Age of the Technological Revolution. Prior to the Industrial Revolution, hand labor and manual work of the skilled craftsman were required to produce quality products. The development of machinery resulted in a reduction of the number of persons engaged in skilled handicraft and a corresponding increase in the use of less skilled labor. The technology of today will require skills over and above either dexterity of hand or physical stamina. Today we need skilled technologists.

COLLECTIVE BARGAINING

The worker at the beginning of the Industrial Revolution who did not have a specifically sought-after skill was at a disadvantage. The result was that trade unions grew rapidly as a method in which labor could bargain collectively by controlling its own supply and establishing work standards and qualifications for various industrial activities.

Organizations that might be classified as labor unions in the United States initially appeared shortly after the American Revolution in certain areas of the newly established country. Organizations having a national character did not appear until after 1825. The first organization to attain any national importance was the Knights of Labor, founded in 1869. At the height of its activity it had approximately 700,000 members. The tendency of labor to band together to attain a better bargaining position was naturally opposed by management, so the Knights, being the first significant organization of its kind, was unpopular and unsuccessful in reaching its aims. It was dissolved in the 1890s. A second organization, formed shortly after the Knights, was more successful. This was the American Federation of Labor (AFL) founded by Samuel Gompers. The original composition of the AFL consisted of unions of skilled workmen.

It was in the 1930s that many labor leaders began to urge workers not in the skilled trades to organize into unions and improve their bargaining status. This led to the formation in 1938 of the Congress of Industrial Organizations (CIO). It was organized on an industry basis and encompassed both skilled and unskilled workers. A competition for members resulted between the two organizations.

Prior to the 1930s, each state had its own laws governing the relationship between management and union-organized workers. Because of the wide diversity of laws in the various states, the Congress, in 1935, passed the National Labor Relations Act. The objective of the Act was to safeguard laborers' rights to bargain collectively. A National Labor Relations Board was created with the right to intervene in disputes between employers and employees and to determine the appropriate union in matters regarding jurisdictional disputes against competing unions. The basis upon which the National Labor Relations Act had been established was the series of incidents that resulted in passage of the Norris-LaGuardia Act of 1932. This 1932 legislative enactment restricted the right of the federal courts to issue injunctions that would, in effect, prevent workers from organizing for collective bargaining.

The National Labor Relations Act of 1935, also known as the Wagner Act, was amended in 1947 by the Taft-Hartley Act. More commonly known as the Labor-Management Relations Act, this amendment was the result of complaints that the Wagner Act, aimed at correcting management abuses, was being unfairly administered by a pro-labor board and had resulted in abuses. The Taft-Hartley Act sought to legislate against unfair labor practices as well as unfair management practices and required unions to file reports on money collected from members and its disposition. This Act was criticized, in turn, as being unduly severe.

In 1955, the AFL and CIO joined forces. The members of the single union thus formed (AFL-CIO) now comprise over a fourth of the working population in this country. There are also several independent unions of considerable size as well as a number of smaller independent unions.

The year 1955 also brought the first Landrum-Griffin Act. The second Landrum-Griffin Act came in 1959. All of these amendments to the National Labor Relations Acts have established regulatory requirements for unions. The regulations governing the holding of elections are completely spelled out in definite terms. Statements must be filed by the unions regarding finances, handling of funds, and the qualifications of union officers. Secondary boycotts, certain kinds of strikes and picketing, as well as certain kinds of labor contracts are prohibited. The government also has been empowered to inspect union activities closely.

In the construction industry union agreements and contracts can be between the national union or the local union on the one side, and an individual employer or an associated group of employers on the other. Union contracts may cover many different aspects regarding working conditions. Both sides must show a willingness to compromise if collective bargaining is to be successful. In the end, however, in accordance with the principles of contract law, the agreement should be acceptable to both parties. Differences may be settled by discussion or negotiation. Independent arbitrators may be appointed if agreement cannot be reached between the parties involved. In cases where the public welfare is involved, an impartial mediator may be appointed by the government, by a governmental agency, or by the courts. The last resort is a work stoppage, or strike.

Among the provisions contained in a union contract are those governing grievance procedures. A grievance is a worker's complaint regarding working conditions. A grievance may concern matters covered by union agreement or it may be a personal matter that is not a part of collective bargaining. A grievance cannot be entered for any matter that has already been included in the body of the union contract. For example, an individual worker cannot request a higher rate of pay than that which the union and the employer have agreed on. However, if he is asked to do a different type of job for which the pay scale is higher at his former pay rate, he has a justifiable grievance. The majority of grievances concern working hazards, the maintenance of tools and equipment, or the attitude of supervisors toward individual workers.

Grievances are handled by channeling them through the lowest level of management through successive steps to the highest. At each step, an attempt is made to settle the difference. Each discussion of a grievance must include representatives of both sides. Since an unresolved grievance may tend to become distorted in its passage from level to level, it is put into writing to provide a record of what has happened at each successive step. Should the two sides not come up to an understanding once the highest level of responsibility has been reached, it may be necessary to call in a neutral outsider who will render a decision.

WORKMEN'S COMPENSATION

Under the common law, an injured workman could recover for his injuries from his employer only by presenting indisputable proof. The entire burden was on the injured party to establish that the injury arose in the course of his employment, that it was the direct result of his employment, and that his employer was negligent. The employer could raise the defense that the employee was guilty of contributory negligence, or that a fellow employee was negligent, or that there was an assumption of risk by the injured party. Recovery under such circumstances was very difficult.

Let us suppose that under the common law, a worker could prove that he was injured during the course of his employment. If he was unable to prove that the employer was negligent in some way, he had no remedy available. If he was able to prove both that he was injured in the course of his employment and that his employer was negligent, should the court find that he, too, was guilty of any negligence that contributed in some way to the accident, he lost. Likewise, if he was able to prove injury in the course of employment, negligence of his employer, and no contributory negligence on his own part, if a fellow employee had been negligent, he would still have no redress. This last burden was known as the "fellow servant rule." In such cases where the work was hazardous to begin with, the law denied relief to the injured workman on the basis that he "voluntarily assumed the risk of his employment" when he agreed to work under such conditions.

Such doctrines as these have now been modified through the adoption of some form of workmen's compensation act regarding industrial accidents.

The workmen's compensation statutes eliminate these defenses. Even the assumption of risk is no longer a defense if it is proven that the injured party was an employee and the injury arose out of and in the course of his employment. The intention of the various statutes is that such injuries be compensated by the relatively uncomplicated procedure of filing a claim without having to resort to outside legal advice.

Certain employers automatically come under the provisions of the Act or the process of court action shall apply. Under one state's workmen's compensation law, the employees who are obliged to participate include, among others, all governmental authorities, the businesses engaged in activities declared extra hazardous, public or common carriers with more than two employees, as well as the following that may be of particular interest to people engaged in the concrete construction field:

> "The erection, maintaining, removing, remodeling, altering, or demolishing of any structure, . . . "
>
> "Any enterprise, business or work in connection with the layout or improvement of subdivisions of tracts of land."

Provisions are made within the act for the waiving of its protection by either a nonobligatory employer or an employee, or both, if the Industrial Commission is correctly notified of such intention. If the employer is the one who elects not to operate under the act, then an injured employee has to resort to the courts for the assessment of the employer's liability. However, if the employer does not waive the provisions of the act, such compensation will be awarded to the employee as provided for by the amounts set out in the body of the act following the decision of the Industrial Commission. The employer is relieved thereby of any liability in a court of law, and settlement is reached by means of the provisions of the act itself. In a court of law, the employer wishing protection is not able to use the old common law defenses.

If the employee elects not to come under the provisions of the act, and attempts to obtain redress through a court of law, the employer is then enabled to use the common-law defenses of contributory negligence, the fellow servant rule, or the assumption of risk.

If both waive the protection of the act, it is as though the employer alone had rejected its protection. He would in such an instance be unable to use the common-law defenses.

The workmen's compensation acts set out the procedures to be followed, reports to be filed, and bond required from the employer.

The employee, on the other hand, is obligated to notify his employer in writing of any disability occasioned in the line of his employment. There will be no compensation for self-inflicted accident or death. Should an employee at any time refuse to submit to or obstruct the medical examination of a competent physician as required by the Commission, his rights to compensation under the act will cease until such refusal has been remedied. A partially disabled employee refusing employment suitable to his disability offered by his employer will not be entitled to any compensation during the time of such refusal.

Workmen's compensation benefits include medical and surgical care following the injury and compensation, as set out in the various acts, based on the employee's wage scales, and the extent of injury or incapacity. Injuries may be temporary or permanent, and resulting disability may be total or partial.

INSURANCE

Consider the case of the average contractor. Almost all of his personal assets are invested in his business. His fixed assets are his offices, storage areas, and equipment. His current liabilities are his expenditures for the materials and the pay that he owes to his workers. There are other expenditures as well, such as utilities, taxes, and equipment maintenance. His profit is derived from the amounts that are paid him after the whole or specified parts of the construction are completed and must reimburse him for the expenditures made as well as give him a profit. The profit attained through the completion of one job must as well provide him with sufficient financial status to obtain materials and employ workmen for the next job.

An accident resulting in injury to himself, his employees, or third parties may well leave him destitute after payments for repairs, hospital bills, or legal claims. The purpose of insurance is to provide protection against this.

The theory underlying an insurance company is that out of a large number of people paying premiums for protection, only a few will have to be reimbursed. Loss visited upon any one of the large number is, in effect, absorbed by all.

Insurance was a development of the law merchant. The famous Lloyd's of London was originally a coffeehouse frequented by seafarers and merchants. When a ship was ready to depart from London, a slip of paper describing the ship, its cargo, the name of the master, and other pertinent information would be passed around among the various customers. Based upon the fact that the safe return of the ship to London after its voyage with a new cargo in its hold would guarantee a profit, those who would be willing to stand liable for a portion of the value in case of loss would sign their names under the description followed by the percent that they would guarantee. If the ship returned safely, each guarantor was paid a share of the profits equal to the share of the responsibility he had pledged.

Today, insurance can be obtained for almost any form of legitimate risk. Contractors as well as architects and engineers can insure against failure to perform a contract for causes beyond their control.

The parties to an insurance contract are the ones who agree (for a consideration) to indemnify a party (this is the insurer, or underwriter) and the one who wishes to protect himself (the insured). The consideration itself is called the premium. The terms are put into writing and are known as the insurance policy.

An insurance policy is a contract. However, further qualifications must be made about the subject matter of this kind of contract. The nature of the risk assumed must be valid; otherwise the contract is void. An insurance policy to cover fines assessed for breaking the law would not be valid risk. Not only must the risk be valid, but the risk or risks assumed must be specified in the contract.

Recovery under an insurance policy can be made only by persons having an insurable interest in the subject matter of the policy. A person having an insurable interest is the one who would bear the loss should the risk specified in the policy occur. Such insurable interest must exist at the time the policy is taken out. In the case of property insurance, the insurable interest must also exist at the time of the loss.

Many states have adopted standard forms of insurance contracts, and prohibit an insurance company from doing business unless the policies they issue contain the standard provisions, and a copy of each form of policy is filed with the proper state official. Necessarily, the insurance contract must be in writing since there is no standard form of oral contract, nor can an oral contract be "filed."

The terms "avoidance" and "forfeiture" have a different meaning when applied to the insurance contract than they do with other forms of contracts. Avoidance means the act of resisting a claim under an insurance policy. Forfeiture is the act of declaring an insurance policy void, either before or after a loss.

Avoidance or forfeiture may result from a defect in the essentials of the contract, an improper form of contract (as when a provision required by a statute is lacking from the body of the contract), or the breach of some warranty or representation. A warranty, according to insurance law, is a statement of fact in the insurance policy that must be true, since the factual material contained in a policy is relevant to the risk that the insurer is willing to assume. For example, an insurance policy covering operating machinery will contain warranties made by the insured as to the age, purchase price, and operating condition.

A representation is an oral or written statement made at or before the closing of the contract but not usually made a part of the subject matter. In order to void the policy, a misrepresentation must be material to the risk assumed. A statement is material to a risk if it induces the insurer to accept the risk. A representation need not be literally true, only substantially true. Innocent misrepresentations, although made without intent to deceive, will still void a policy if they are both material and untrue.

In many states the distinction between a warranty and a representation has been abandoned, and all statements made in connection with insurance are considered to be representations. As a result of such a decision, an insurance company wishing to resist payment on a policy has the burden of proof that the statement was material or, if not material, made with the intent to defraud.

SURETYSHIP

A surety is one who will guarantee the integrity of another person. Suretyship, therefore, is the obligation to answer for the debt, default, or tortious actions of another person. Debt concerns money obligations, default concerns contract obligations, and tortious actions concern noncontractual obligations. Suretyship can be established by the giving either of property (real surety), or a personal promise (personal surety). Generally, the term "suretyship" refers to personal suretyship, the promise of some person other than the debtor to make good the debt in case the debtor defaults.

Suretyship always arises out of a contract. It is, in effect, a second contractual arrangement that develops from a prior contractual arrangement. The first contract is the principal or primary obligation. The second contract, the suretyship itself, is the secondary or collateral obligation. The person who is liable for the primary obligation is the principal debtor. The person to whom the principal debtor is obligated is the creditor. The person obligated under the secondary contract is the surety, and he is also obligated to the creditor. Suretyship, as such, implies the existence of two contracts and three contracting parties.

Although suretyship and guaranty are frequently considered to be one and the same, there are several distinctions that can be drawn. For one thing, the obligation of the surety is in a given amount and is entered into at the same time that the principal debtor enters into the primary obligation. The obligation of the guarantor begins at the time the principal debtor defaults on his primary obligation and is for whatever sum the principal debtor has left outstanding on his obligation. A surety is liable without notice of the principal debtor's default, while a guarantor is entitled to receive notice of the principal debtor's default.

As both suretyship and guaranty are contractual obligations, the principles of contract law apply. As both constitute promises of one party to be liable for the debts of another, both of these contractual obligations must be in writing.

Let us quickly list the types of guaranties that may be found in the course of business transactions:

A *guaranty of collection* is one in which the guarantor agrees to pay if the principal debtor cannot. It must be shown by the creditor that the principal debtor cannot pay by having exhausted all other legal remedies to make him pay.

A *guaranty of payment* is one in which the guarantor agrees to pay if the principal debtor does not and the guarantor is liable immediately on the principal debtor's default.

An *absolute guaranty* is a definite promise to make good another's obligation, without qualification.

A *conditional guaranty* is one in which the obligation to answer for another's default is made to depend on some condition.

A *single guaranty* is one limited to a given transaction.

A *continuing guaranty* is one that is not limited to a single transaction but covers a succession of liabilities for which, as and when they accrue, the guarantor becomes liable; such a continuing guaranty usually contains no fixed time limit, although there may be a fixed money limit.

A *general guaranty* is addressed to no particular person but to the public at large and may be enforced by anyone to whom it is presented and who acts on it.

A *special guaranty* is one addressed to a particular person who alone can take advantage of it and to whom the guarantor will be responsible.

A surety may be relieved of liability through the principal contract being void under the principles of contract law. The surety is also relieved from liability when the principal contract is discharged. A third way in which the surety may be relieved from liability is when his risk has been materially altered without his consent. Such variations can occur through the adding or substitution of new debtors to the principal contract; the changing of the duties of the principal debtor under the principal contract; the changing of any material term of the principal contract such as the amount, place, time, or manner of making payments; or by the creditor's surrender of security that has been given to him by the principal debtor in connection with the principal obligation.

A surety is not discharged even when the creditor has delayed proceeding against a principal debtor who has defaulted. To remedy this situation some states have enacted statutes requiring that a creditor proceed against the debtor within a given period of time. If the creditor does not comply with this requirement, the surety is discharged. This applies in those instances where the creditor must exhaust his remedies at law before holding the surety liable.

The losses that a surety may sustain in taking over a debtor's obligation can be remedied in three ways.

If the surety has discharged the debtor's obligation and the creditor is in possession of property that the debtor used to secure the loan, the surety may require the creditor to turn over the security. The claim of the creditor is then the property of the surety and the surety becomes creditor by subrogation. In order to repossess the property, the debtor may pay off his obligation to the surety.

Having paid the debtor's obligations, the surety may then have a contract implied by law that the debtor must indemnify the surety for his losses. The surety's right created by this obligation—known as indemnification—does not require the turning over of any security as does subrogation. However, both indemnification and subrogation give the surety the right to proceed at law against the debtor.

If there is more than one surety, and one pays the complete obligation, he may proceed against the other sureties to make good their part of the suretyship by the means of the remedy of contribution.

SURETYSHIP IN CONSTRUCTION

The concept of suretyship is important to the construction industry because it provides assurance to all parties concerned, from general contractors to developers and lenders, that a certain project or part thereof will be completed. Surety bonds are an integral element in the financing of most construction projects, with bonding entering the picture in different forms. The general distinction between types of bonds is given below.

1. Completion Bond

A completion bond guarantees to the obligee that the project will be completed and paid for. It includes in addition to an obligation to complete the job, assurance of the financial responsibility of the owner or the person responsible for the completion of payment. Completion bonds are expensive, rare, and in many cases unobtainable. The word completion bond is used rather frequently when what is really meant is a performance bond.

2. Performance Bond

A performance bond obligates the contractor to complete the work for the amount to be paid in accordance with the contract. The surety agrees in the event of a default that the surety will complete the work.

3. Payment Bond

Payment bonds are bonds that obligate the general contractor to pay subcontractors and to keep the property free and clear of liens. It should be noted that on some forms of payment bond, although the owner is protected against loss for failure to pay, there is no direct obligation to the subcontractor involved. Consequently, although the owner may be protected, there is no assurance that the subcontractor will be paid.

4. Bid Bond

A bid bond is customarily required in an amount equal to a set proportion of a bid price. It assures that if the bid is approved, the bidder will enter into a contract in accordance with his bid. A contractor applying for such bond should be aware that the bonding company may approve his bid bond, but might not approve his performance bond. This would make it impossible for him to complete his contract obligations. Therefore, when obtaining bid bonds, it is suggested that the contractor also make sure that he will be accepted for the performance bond.

5. Lien Bond

Another bond known as a lien bond is a bond in which the only protection given to the owner is that no liens will result from the work. This bond does not guarantee that the work will be completed, started, or paid for.

CHAPTER 8

LAW OF TORTS

In the legal aspects of the concrete construction industry, a violation of a personal or individual right or duty constitutes a civil wrong. Where the violation is one of the "natural" rights of an individual, it is a civil wrong known as a "tort." Natural rights are personal liberty, personal security, or property.

Assault, battery, libel, slander, and false imprisonment are torts. One's property can be affected by wrongs giving rise to the torts of trespass, conversion, nuisance, and "waste." There are torts that bear relation to business transactions, which are often referred to as "business wrongs." Some examples of this type of tort are fraud, deceit, and wrongful interference with a contract. There is also a large group of torts that are based on negligence. These may affect an individual's person or property or both.

TORTS BASED ON NEGLIGENCE

Negligence is the failure to use reasonable care. Negligence is the cause of many torts that arise when working in the field of concrete construction.

In determining whether there has been negligence, the courts will use the yardstick of "the reasonably prudent man." In the case of a tort of negligence, the infliction of an injury is completely unintentional. Therefore, if the conduct of the accused is such that an ordinarily prudent person faced with the same set of circumstances would have done exactly the same thing, there has been no negligence on the part of the accused.

The diligence that will be exercised by a reasonably prudent person will vary with the circumstances. The driver of a truck carrying explosives or flammable liquids can be expected to be more careful than the driver of a car. The standard of care that is proper in any situation is dependent upon the nature of the surrounding circumstances.

Let us take an actual case to see how the court will treat the concept of negligence. The construction project was an addition to a high school and certain remodeling work was involved in making the addition. The parties are the contractor's injured employees and the architectural firm.

The architectural firm had entered into an owner-architect agreement with the Community Unit School District to design the addition. The agreement, in manuscript form, contained no unusual contract provisions. However, in the light of the circumstances surrounding the accident upon which the case is based, special attention should be paid to the wording of the provision as to "supervision." This section of the agreement read as follows:

> "The architect will endeavor to guard the owner against defects and deficiencies in the work of contractors, but he does not guarantee the performance of their contracts. The supervision of an architect is to be distinguished from the continuous superintendence to be obtained by the employment of a clerk of the works."

The contract awarded to the general contractor was in the amount of approximately $200,000. The construction required that the contractor shore the existing gym roof and remove one of the gym walls. Tubular steel scaffolding was used in the shoring. The shoring was installed to support each beam and truss of the existing roof. A four-legged scaffold was located under each truss with the load being equally distributed on each leg. No stress was noticed on the shoring; nor was there any other indication that the shoring was overloaded during the progress of the work.

All work connected with the positioning, construction, and maintenance of the scaffolding and shoring, the temporary roof supports, the removal of existing supports and columns was performed under the direction and control of the contractor. Neither the school district nor the architect exercised any control over such work.

On May 3 three of the contractor's workmen were in the process of removing the last column supporting the gym roof. The procedure being used to remove the columns was as follows. A crane cable was attached to the column to be removed. One workman on the ground would cut the column with a torch, and two workmen on the roof would loosen the bolts. The crane operator would then pick up the column.

The workman on the ground had cut through the column at the base, and one of the workmen on the roof had removed what he thought was the last bolt. The crane cable was tightened to remove the last column but it would not budge. It was at this point that the worker on the roof discovered another bolt. He had just removed this bolt when the roof and shoring collapsed. All three of the contractor's employees were seriously injured.

Subsequently, the injured workmen filed suit to recover damages for the injuries they had sustained against both the architect and the school district. The plaintiffs' contention was that the defendants had negligently failed to provide the plaintiffs with a safe place to work or with adequate support or shoring to the gym roof, and particularly that the architects had been negligent in their application of the degree of skill that would customarily be required of competent architects in connection with a project of this type. The architects, in turn, filed a third-party complaint against the general contractor.

Over the architects' objections, the actions involving the school district and the contractor were dismissed, and the action involving the architects went before a jury in September three years later.

Pretrial investigation and testimony at the trial failed to establish the exact cause of the collapse. However, it was determined after the collapse that the load of the existing roof was well within the manufacturer's specifications for the tubular steel scaffolding that was being used.

The basis of the plaintiffs' case was that there was insufficient information on the plans to enable the contractor to compute the load factor on the existing roof. The architects should have computed this load factor and supplied the information to the contractor. The architects had received a fee for supervision as well and had done nothing to carry out the duties to which they were obligated by the payment of such fee.

The architects' defense was to refute the plaintiffs' contentions and to establish that they were not liable under the provisions of the Illinois Structural Work Act. Expert testimony was offered to the effect that there was sufficient information on the plans to calculate the load factors. Admissions were made by the architects that they did not make such calculations or furnish them to the contractor. Evidence was also brought forward to show that the load factor of the shoring was more than adequate to support the existing gym roof. The architects' contention, backed by evidence, was that architectural supervision under the circumstances of the case amounts only to such observations as are necessary to assure that the owner receives a building that is constructed in accordance with the plans and specifications. It was further contended that the shoring was what is termed a construction procedure and that an architect has no right or obligation to interfere with the techniques or methods of construction procedures. Furthermore, the "shoring" employed in the construction was not a "scaffold" as defined in the Illinois Structural Work Act, and the liability set forth for such a scaffold would not apply in this case. (The liability referred to attaches to a party under this provision not only when a dangerous condition is known to him, but also if by the exercise of reasonable care, he could have discovered the existence of such a dangerous condition.)

The jury found the defendant liable and judgments were rendered for the plaintiffs in the amount of $125,000. The finding of the court was that the architects were negligent in failing to provide adequate support for the existing roof and had failed to exercise the degree of care that an architect should exercise.

When the case was appealed to the Appellate Court, that court held that the temporary shoring constituted "supports," "Stays" or "other mechanical contrivances" within the Illinois Structural Work Act.

The court also held that for the liability as provided for by this Act to attach to a party, it is not necessary for a reckless disregard of the Act's provisions to be proven.

The decision was reviewed by the Illinois Supreme Court. In the opinion handed down, particular significance was attached to the status given the architect by the terms of the contract awarded to the contractor. Article 38 of the contractor's construction contract read as follows:

> "The Architect shall have general supervision and direction of the work. He is the agent of the Owner only to the extent provided in the Contract Documents, and, when in special instances, he is authorized by the Owner so to act, and in such instance he shall, upon request, show the Contractor written authority. He has authority to stop the work whenever such stoppage may be necessary to ensure the proper execution of the Contract."
>
> "As the Architect is, in the first instance, the interpreter of the conditions

of the Contract and the judge of its performance, he shall side neither with the Owner nor with the Contractor, shall use his powers under the Contract to enforce its faithful performance by both."

In the opinion of the Illinois Supreme Court, the parties had agreed that the architects must exercise reasonable care in the performance of their duties. A lack of such reasonable care would make them liable to persons who may foreseeably be injured by such failure regardless of whether or not a relationship had been established between the architect and the injured party by contractual means.

The court felt that the principal question was the extent of the architects' duties. The evidence, as presented, showed that the architects did not prepare detailed specifications for the temporary shoring, did not compute or indicate on the plans the load that would be placed upon the shores or provide the contractor with a safety factor to be used in the shoring, and did not oversee and inspect the shoring as used. The record can be viewed as pointing to negligence on the part of the architects based upon one or more of these omissions.

The architects' contention that the general duty to supervise the work was merely to see that the building as constructed was in conformance with the plans and specifications cannot be upheld in the circumstances of the case. Although the method or technique of construction is said by the architects to be a matter over which they had no control, they did have a right to intervene or even stop the work if the contractor began to shore in an unsafe or hazardous manner in violation of the construction contract. The wording of the construction contract that would make such negligent shoring a violation is contained in paragraph 12 of that contract. It states that the contractor, "shall take all necessary precautions for the safety of employees on the work" and includes in such precautions "all bracing, shoring and sheeting as required for safety and for the proper execution of the work. . . ."

The Court, therefore, concluded that if the architects knew, or in the exercise of reasonable care should have known, that the shoring was inadequate and unsafe, they had the right and corresponding duty to stop the work until the unsafe condition had been remedied. By breach of such duty, the architects are liable to those plaintiffs who could foreseeably be injured by the breach. It was also held that the wording of the Illinois Structural Work Act regarding persons "having charge" is applicable to the architects as their right to stop the work if it was being done in a dangerous manner, thus placing them in a position of having charge.

It is true that the dissenting opinion did not find such a "duty" as set forth by the majority of the Court. This divergence of opinion is based upon the fact that the architects' contract is more or less a standard form. The standard wording that the architects "had the right" to insist upon a safe and adequate method does not make "a duty."*

It must be noted that the current trend of the courts throughout this country has been to extend the areas of liability to third parties by lines of reasoning that do tend to make a "right" or an "option" equivalent to a duty.

THE REASONABLY PRUDENT MAN

In attempting to define the standard of care of a reasonably prudent man, the statement has been made by the courts, "Negligence is the omission to do something which a reasonable man, guided upon those considerations which ordinarily regulate the conduct of human affairs, would do or something which a prudent and reasonable man would not do." Would a reasonable, prudent man have been obligated to do something to prevent the following accident from happening?

Workmen employed by a roofer were engaged in putting a new roof on an apartment house. In the backyard of the tenement house was a garage also belonging to the building owner but leased to other people who were not tenants. The garage had a flat roof some 6 ft above the ground. The owner of the building had given the roofers permission to place their tar kettle in the yard while they were working, and they had put it about 3 ft from the wall of the garage. The tar kettle was a cylinder 3 ft high with a fire pot and a container in which tar was kept hot. On the day of the accident, the workmen had finished with the hot tar and had extinguished the fire shortly before noon. The top of the tar kettle had been covered with a piece of sheet metal over which was placed a slab of wood. About 10 or 12 in. of tar remained in the pot.

Late in the afternoon, while all of the workmen were on the roof, the six-year-old son of one of the tenants of the building was given a penny by a 10-year-old neighbor boy to jump from the roof onto the cover of the tar kettle. In doing so, he hit the cover on the side. The cover slid off and the child fell in. He subsequently died as the result of the burns he suffered. The lower court found for the plaintiff, the father of the child, and declared that the defendants, the roofer and his workmen, were

* Miller et.al v. DeWitt-Amdal & Associates et.al in the Supreme Court of the State of Illinois (1967), as discussed in *Inland Architect*, September, 1967.

negligent. This question was presented to the higher court; "Was there a violation of any duty of care which the defendants owed to the deceased?"* The court held that due care does not require that special precautions be taken against that which is only remotely possible. The fact that the fire was out and the kettle covered makes it impossible that a child could get into the kettle from the surface of the yard without intentionally intending to do so. Although there was evidence that sometimes children did jump from the garage roof, their presence there was actually that of trespassers. Whatever duty the defendants owed to children present in the yard, they were not bound to anticipate and to guard against the bare possibility of children jumping through the cover into the pot from an adjoining roof.

STANDARD OF CARE

However, it is possible that the standard of care involved in the presence or absence of negligence can be set forth by a statute. In a case based upon the plaintiff's falling through an unguarded elevator shaft, there was a statute in existence that required all machinery, shafting, belting, etc. in factories and workshops when placed in such a position as to be dangerous to employees to be securely guarded when practicable. In another section, all hoistways and elevators were required to be protected by trapdoors or guardrails. The final section imposed a penalty for the violation of any of the provisions of the act. The plaintiff was a municipal fireman called in to help extinguish a fire that had started at a time when the factory was not operating. The only class of persons named as being specifically protected by the statute were employees, so neither firemen nor any other classes of persons came within the statutory provisions. Was the factory owner liable?

The court held that for liability to attach, it would be necessary for the plaintiff to show that he was one of those intended to be protected by the statute. In an action based upon a neglect of duty, it is not enough for the plaintiff to show that the defendant neglected to perform a duty imposed by statute for the benefit of a third person and that he would not have been injured if the duty had been performed.†

The common law imposes a duty of average care on anyone who undertakes to perform a service, whether he is or is not compensated for his

* Marengo v. Roy, Supreme Judicial Court of Massachusetts, 1945. 318 Mass. 719, 63 N.E. 2nd 893.

† Kelly v. Henry Muhs Co., Supreme Court of New Jersey, 1904. 71 N. J.L. 358, 59 Atl. 23.

services, with regard to all people with whom he comes in contact while performing that service. Many statutes and ordinances are so worded as to express a universally obligatory rule of conduct. The general duty is established by the statute, and a deviation from such general duty without good cause is a wrong. A wrongdoer is responsible for the damages resulting from his actions. Failure to perform a statutory duty indicates that negligence might exist. If the wording of a statute does not specify the people to whom the protection is to be provided, the absence of the necessary precautions would satisfy the common-law requirement of average care.

RES IPSA LOQUITUR

There is a concept of the law regarding the proof of negligence that is termed in Latin "res ipsa loquitur." The meaning implied by this term is that an occurrence could not have happened if there had not been negligence. If the article causing the injury was, at the time of the injury, in the possession and exclusive control of the defendant, and the occurrence that took place was such that it would not happen in the ordinary course of events if the one having control had used proper care, res ipsa loquitur would apply. For example, a piece of heavy equipment will not roll off an incline if adequately blocked. If it does, the injured party would have a good case for negligence. However, if the injured party had removed blocks causing it to roll off the incline, it is his own negligence and the blame cannot be placed on anyone else.

SINE QUA NON

The "sine qua non" rule in the law refers to the actual cause from which the injury resulted. If the defendant had not committed the act for which he is being held responsible, there would have been no injury being done to another person. The fact that there is a possibility that the act of the defendant caused the injury is not enough to attach liability; there must be a direct causal relationship.

The erection of a defective structure that collapses and causes an injury to a third party has such a causal relationship. Without the defect, the structure would not have fallen. If the defendant contractor had not supplied the defective structure, the third party would not have been injured.

Now let us suppose that the failure has been caused by material obtained in a defective condition from a supplier. In this instance, the law of restitution applies.

(1) Where a person has supplied to another a chattel which because of the supplier's negligence or other fault is dangerously defective for the use for which it is supplied and both have become liable in tort to a third person injured by such use, the supplier is under a duty to indemnify the other for expenditures properly made in discharge of the claim of the third person, if the other used or disposed of the chattel in reliance upon the supplier's care and if, as between the two, such reliance was justifiable.

(2) The rule stated in Subsection (1) applies where a person has negligently made repairs or improvements upon the land or chattels of another whereby both become liable in tort to a third person.

PROXIMATE CAUSE

Not all instances of injury can be pinpointed with exactitude. Consider, for example, the collapse of a hotel in Basel, Switzerland, which occurred in 1901. The collapse involved one wing of a five-story beam-and-slab structure. The design of the structure was prepared by a Parisian firm that merely supplied the plans and undertook no control of the fieldwork. The local building department in Basel undertook only such procedural control as was strictly prescribed by law. Finally, the contractor in charge of the job let matters take their course with little organization.

The structural plan required that masonry pillars were to rest on concrete footings at basement level. These pillars would be the support for the beams of the first floor. On the day of the collapse, the pillars had not been built and the first floor beam was held up by wooden supports. The construction had progressed to the point where work was to start on the pillars. In order to lag the masonry more conveniently, the temporary wooden supports were removed. The columns on the first floor, therefore, were unsupported except by the floor beam. Two hours after the removal of the temporary supports, the building collapsed.

Was the removal of the temporary supports the one event that brought about the collapse? If the supports had not been removed, would the building have remained standing? Or to ask this question in a legal manner, "Was the wrongful act the proximate cause?" The investigators also found that sand and gravel taken from the excavation had been used in the concrete, that the bottom of the columns were poorly consolidated, that no testing was done of the cement or concrete, and that there was no organization among the various subcontractors, as well as no undue haste in the construction work.

It has been suggested that in determining proximate cause, a circle be imagined with the accident itself as the central point. All factors that would contribute to the accident can be placed at such distance from the central point as their foreseeability of causing an accident can be determined. The factor located nearest the central point is not necessarily the proximate cause of every injury resulting from an accident, but it may be the proximate cause of the accident itself. The circle of liability may encompass many different claims. The proximate cause of such claims may be time, place, or location of the injured party. Such a placement of the various causative factors around the central event of the accident can result in a proximate cause for each type of injury done being obtained. For example, it is foreseeable that excavation for a foundation could cause the adjacent subsoil structure to change. It is also foreseeable that a shift in subsoil structure could cause a strain on a building adjacent to the excavation. What is the degree of foreseeability that the building would shift and crack the sidewalk, and that someone walking by would stumble over the crack, fall, and be injured? Would the degree of foreseeability be changed if it were known that the locality had a relatively unstable subsoil?

The rule has been established that if subsequent to the original act of negligence, a new cause has intervened that is of itself sufficient to stand as the cause of the injury, the original negligence is too remote to be considered the proximate cause. If, however, the new cause could have reasonably been foreseen to have resulted from the original negligent act, the chain of causation is not broken by the intervention and the one responsible for the original negligent act still is liable for the injury.

CHAPTER 9

THE INJURED AND CONTRIBUTORY NEGLIGENCE

One of the common-law defenses that an employer had in a tort involving injury to a worker was that the worker himself had contributed to the injury. Another defense was that a fellow worker had been negligent and had caused the injury to the worker. A third was that the worker himself had assumed the risk when he took the job.

WANTON AND WILLFUL MISCONDUCT

In some of the old legal actions in tort law, the language used concerning contributory negligence is that if one person is at fault, it does not excuse another for not using ordinary care in his own protection. When the wrongful action that brings about the injury is a willful tort such as assault and battery, contributory negligence is clearly not a defense for the wrongdoing party. This is also true of reckless or wanton misconduct. Negligence consists of inattention or inadvertence. Wantonness is a situation where it is realized that danger exists through one's actions, but the welfare of other persons is recklessly disregarded. The implication is that there is a willingness to inflict injury. The intent or willingness to inflict injury is the basis of many of the criminal torts.

CONTRIBUTORY RECKLESSNESS

A laborer was employed by a construction company on a project for the reconstruction of a highway. The two eastbound lanes on which work was progressing had been closed to traffic by means of a barricade. Workmen were engaged in various duties in the barricaded area. The laborer in question was sent by his foreman to the supply shed that was located west of the barricade on the north side of the highway. The westbound portion of the highway was comprised of two lanes, which, because of the construction work, had been converted to two-way traffic. The laborer crossed the two traffic lanes and picked up the hand tools requested by his foreman without incident. After recrossing the two lanes open to traffic, he proceeded to walk along the south edge of the lane open to westbound traffic next to the area under construction. He was struck from behind by a speeding car and subsequently died. The driver's view was unobstructed and he had been aware of the victim walking ahead of him for more than 200 yds. A nearby workman testified that he did not hear the brakes being applied until after the victim had been struck. The driver contended that the victim had been contributorily negligent in walking in close proximity to moving traffic with his back to approaching cars. Will the defense of contributory negligence hold up? Is the driver guilty of wanton negligence? Why?

Let us change our facts a little. Suppose, the driver, having noticed the workman walking along the side of the road, had slowed down. Approaching traffic prevented his driving around the pedestrian. The workman was aware that a car slowed down to a snail's pace behind him and looking back, deliberately moved more directly in front of the car. The driver honked his horn impatiently and the workman made a motion that he was moving out of the way to the side of the road. As the driver stepped on the accelerator to move around him, the workman deliberately stepped in front of the car. Once again the driver pleads contributory negligence. In this instance, the workman was guilty of contributory recklessness. Contributory recklessness is a bar to reckless negligence. Contributory recklessness is a reckless action one undertakes being fully aware that injury might result.

ASSUMPTION OF RISK

The assumption of risk is also an action that one undertakes being fully aware of its danger. Assumption of risk can be used in cases other than the performance of a hazardous occupation. If, for example, a person exercising due care in driving a vehicle, undertakes to drive around a barricade after having noted a sign stating "Travel at your own risk," he has assumed the risk of damage to his automobile or injury to himself. However, assumption of risk does not apply where a person is placed, through the negligence of another, in such a position that he is compelled to choose, in the face of apparent danger, between two hazards. For example, a workman on a structure that begins to give away has a choice of jumping to a lower level or moving to an adjoining building where moving machinery is positioned. Whichever action he chooses to take, he is not held to have assumed the risk in case of injury. If it is brought out in court that had he made the other choice no injury would have resulted, this is not considered a defense, since he was not afforded the time to carefully weigh all the aspects of the situation, as would be expected of a reasonably prudent man.

STATUTORY VIOLATIONS

If there is a statute designed that has the objective of the public safety in mind, and the plaintiff received his injury as a result of the defendant's violation of that statute, the defendant is, of course, liable for the injuries. But what if the injured party were contributorily negligent? Even if this were the case, the defendant would be liable. The restatement of torts brings out that if the defendant's negligence consists of the violation of a statute enacted to protect a class of persons, herein the general public, from their inability to exercise self-protective care, a person so protected is not barred by his contributory negligence from recovery for bodily harm caused by the violation of the statute.

LAST CLEAR CHANCE TO AVOID DANGER

A situation may arise in which the defendant claims the plaintiff was contributorily negligent, and the plaintiff contends that he could not have been contributorily negligent since he was unaware of his danger. The elements of such a situation are that the defendant is aware of the plaintiff's presence and possible danger, realizes that the plaintiff is inattentive and unlikely to discover his danger in time to avoid harm, and then is negligent in using reasonable care to avoid harming the plaintiff. For example, a workman standing next to a pile of debris has his back to the crane operator, who is in the process of dumping material on the pile.

The workman does not turn or give any sign that he is aware of the material about to be dumped. The crane operator swings the crane out over the workman's head, a loose plank falls and strikes the workman, injuring him. The plaintiff will be able to recover (although he was negligently inattentive), since the defendant was aware that the worker had not discovered his peril.

Suppose the workman had gone to the pile of debris against instructions. He realizes, too late, that material is falling from the crane. The defendant either knows that the plaintiff is in helpless peril or has reason to believe that even if the plaintiff would attempt to help himself, he could not remove himself from danger in time. The possibility exists that the defendant could have avoided putting the plaintiff in helpless peril if he acted with the vigilance it was his duty to exercise. If the defendant fails to use preventive measures left to him to avoid hurting the plaintiff, the plaintiff will recover for his injuries.

IMPUTED NEGLIGENCE

Negligence can be imputed to a third party who is neither the plaintiff nor the defendant in a tort action. Consider this example. Sidewalk traffic had been diverted around the construction area of a downtown office building by means of a boardwalk around the site that extended into the street. Through negligence on the part of the contractor's workmen, one of the boards was not nailed down correctly. A mother and her small child were walking on the boards and when she stepped on the loose board, it flew up, struck the child just below the knee, and caused an injury. The mother did not seek medical attention for the child, but subsequently the injury caused permanent crippling damage. The defendant contractor contended that contributory negligence be imputed to the child through her mother's negligence. In some of the states, the negligence of the mother will be imputed to the child. In others, it will not.

There are instances where imputed negligence can arise through a "joint enterprise." The joint enterprise may be likened to four men leaving together for a hunting trip. One owns the car, and the other three will share equally in the driving and the expenses incurred. Let us consider some of the types of accidents our vacationers could have.

The owner of the car is driving the car. The car overturns by the driver's negligence and his three passengers receive injuries. Is the owner liable? Yes. To all three passengers.

One of the three passengers is driving the car negligently. The car turns over. The owner and the other two passengers are injured. Is the driver liable? Yes. To the owner and the two passengers. Is the owner liable? No, they are members of a joint enterprise.

The owner is driving. The car negligently hits a third party. Who is liable? The owner and the three passengers. Once again, they are members of a joint enterprise.

The car with our accident-prone joint adventurers is struck by another car. All joint adventurers can hold the third party liable. But suppose one of our joint adventurers is contributorily negligent? Then the contributory negligence is imputed to all of them.

LAW OF TORTS—NUISANCE

So far we have dealt with tortious action involving the area of negligence with regard to the actions of plaintiff and defendant. There are other tortious actions that arise through the plaintiff's property rights as well as the defendant's property rights. These are nuisance, and the rights of invitees, licensee, seller and buyer, tenants, etc.

PUBLIC NUISANCE

Originally, public or "common" nuisance was considered as a type of low-grade crime. The earliest cases of public nuisance involved obstructions on the public highways. This is still a typical public nuisance. The nature of the crime was gradually enlarged to cover other miscellaneous invasions of the public right. In 1536 it was recognized by the English courts that not only could a minor crime against the public be involved, but also tortious conduct against a private individual who had suffered special damage. The obstruction of a navigable stream would constitute a crime as an interference on the public convenience.

PRIVATE NUISANCE

The tortious action of private nuisance developed as an unreasonable interference with the use or enjoyment of a property. The interference might be with the physical condition of the premises, as for example, blasting or vibration that might damage an adjacent house. The health of the property owner could be endangered as by unsanitary conditions adjoining, such as the disruption of a sewer system during excavation. The comfort or convenience of the property owner might be endangered by smoke, odors, noise, or heat. Or the property owner may feel that there is a threat of future injury as in the case of stored explosives.

The original character of private nuisance as an invasion of property use is still effective, and only those who have a property interest in the land can bring an action. A guest, an employee, or a lodger cannot maintain an action of nuisance. Where there has been bodily injury sustained by such a person, although the basis of the action would appear to be nuisance, it is really a tort based upon ordinary negligence or breach of duty.

In fact, few nuisance cases consider whether the tortious action is based upon intent, negligence, or strict liability. Whatever the original conduct may have been, if the defendant, after receiving notice that damage is being done to the property rights of another, continues, he is considered to be committing an intentional tort.

CONTRIBUTORY NEGLIGENCE

Intentional conduct, or willful or wanton conduct, on the part of the defendant will not permit him to use the defense of contributory negligence on the part of the plaintiff as is the case in other tortious actions. However, merely negligent conduct will permit a defense of contributory negligence the same as in other negligence cases. If the interference with the plaintiff's property rights has been unintentional, the plaintiff must prove that negligence existed.

A residence is located near a rock quarry owned by the defendant. From time to time, the defendant, in his quarrying operations, performs blasting operations to dislodge ledges of rock. Evidence exists that actual injury has been done to the house by the blasting. It is admitted by the homeowners that the blasting was a customary sort and not of a more or less temporary nature. The homeowners had given the defendant notice that injury was being done to their property and that the housewife had become ill of a nervous complaint as a result of the noise, and her fear for the condition of her home. It was held that the safety of property in general is superior in right to a particular use of a single piece of property by its owner. Therefore, the defendant's conduct constituted a nuisance. Recovery was also permitted for emotional disturbance without physical contact in a case where a defendant is maintaining a nuisance.*

The chief industry in a certain Massachusetts village was the quarries and had a finishing plant of considerable size near the plaintiff's property.

* Dixon v. New York Trap Rock Corp., Court of Appeals of New York, 1944. 293 N.Y. 509, 58 N.E. 2nd 517.

The defendant's business had been established some 50 years before the legal action. Some 10 years before the court action was instituted, changes in the technology of finishing granite required the defendant to install finishing machines. At the time of the court action, some seven finishing machines were in operation. The machines, being operated by compressed air, produced loud, penetrating, and disagreeable noises. Since dust is produced in the finishing process, which could be harmful to the health of the workmen, the machines are operated out of doors. The plaintiff brought an action claiming the noise interfered with the reasonable comfort and enjoyment of life.

The court held that the test of a nuisance through noise depends upon all the surrounding circumstances. In this instance, the material prosperity of the community depended upon the industry involved. In such an instance, "extreme" or particular rights cannot be enforced that would be a detriment to the general welfare of the community. "One who settles in a district which possesses natural resources of a special kind, cannot prohibit the development of those resources merely because it may interferefere in some degree with personal satisfaction or aesthetic enjoyment."

An industry, proper in one place, might be highly detrimental in another. One cannot move into a factory district and expect quiet. Nor can a factory invade the territory set aside for residences without disturbing the general character of the community.*

DUTIES TO THIRD PARTIES

As can be seen from these discussions, there are duties owed to many people by anyone in the construction field. There are duties that one owes to the people with whom he does business, with whom he works and who work for him, who have property rights, and to third parties with whom his actions might bring him in contact. What would be the extent of duties owed to the people whose presence is unknown or unanticipated?

What is the duty that a contractor owes to an adult trespasser in a construction area? The fact that there is a trespasser on the premises is usually unknown and in most cases, unanticipated. (There are cases, of course, where it is a well-known fact that a certain area has been used as a short-cut from one area to another. In such cases, depending upon

* Stevens v. Rockport Granite Co., Supreme Judicial Court of Massachusetts, 1914. 216 Mass. 486, 104 N.E. 371.

the circumstances, the presence of the trespasser might be anticipated. However, the fact that the trespasser is in the area at all has made him contributorily negligent in being a wrongdoer. He has, therefore, assumed the risk.) One who owns or is in control of land does not have the burden of looking out for the welfare of those who enter without right or permission. If, of course, an artificial hazard has been created, such as a high tension wire, which might cause death or injury to trespassers constantly intruding on a given area, the owner is *under the duty of posting notice* of the hazardous condition. Although no duty of care is owed to the trespasser, once the presence of the trespasser has become known, there is a duty not to exercise such willful and wanton conduct that injury will result.

The law, however, is somewhat different with regard to trespassing children. Children are attracted to dangerous areas. A dangerous condition on strictly private property that might prove fascinating to trespassing children is known as an "attractive nuisance." Although no one is under a duty to provide a safe playground for trespassing children, there is a duty to provide ordinary protection for young trespassers from attractive nuisances. Some localities use the rule of law that the young trespasser must have been attracted to the premises by the condition itself in order for there to be any recovery from an injury caused by an unprotected attractive nuisance. This is, however, the minority opinion. Most follow the rule that the attraction is important only insofar as it might be anticipated to allure a child to trespass.

LICENSEES

A licensee is one who has been granted the privilege of entering a place of business but not in the transaction of business. Once again, as the presence of the licensee might be unknown or unanticipated, there is no duty of care toward the licensee, but there is the duty to refrain from willful and wanton behavior which might cause injury. A guest in one's home may be considered a gratuitous licensee. If, of course, the owner knows of the existence of a dangerous condition, he is under a duty to warn the licensee.

INVITEES

Invitees are those who enter premises on business concerning the owner. The term "business visitors" is in more frequent usage today. As the visitor of this type may or may not mean a financial benefit to the owner, such a visitor is entitled to expect that the possessor will take reasonable care to make him aware of the actual condition of the premises and either take safety measures or warn him of dangerous conditions. Safety glasses for visitors to a factory are this kind of precaution.

PUBLIC SERVANTS IN DISCHARGE OF DUTIES

Not all injuries on the premises put the owner in a position of liability. Consider this seemingly ironclad case of personal injury.

A member of a city fire department had been directed to inspect the defendant's premises following the requirements of a city ordinance. The wording of the ordinance conferred upon him the power and authority of a policeman in the discharge of his duties. While going down a back stairway, he stumbled on a defective tread and fell. The stairway was not open to the public, it was used only by employees. There was nothing in the nature of a trap nor any intentional concealment of the condition. Since the authority conferred on a fire inspector entitles him to enter any building for inspection and provided for punishment by fine or imprisonment of anyone refusing such inspection, he is not an invitee or business visitor. Nor was he present from the discharge of any duty owed to the owner. He was there to discharge a public duty. The owner, therefore, was not under a duty to the fire department to keep the property in a reasonably safe condition. However, there would be a duty of disclosure on the part of the owner as far as any known danger was concerned.

LANDLORDS

If a landlord or owner of premises has retained the privilege of entering for inspection, he is still in control of the premises and liable along with the occupier. A lease which contains a provision for repairs to be made by a landlord will attach legal liability to the landlord in the event of an injury sustained by the tenant through the landlord's failure to repair. Where no damage has resulted, the failure to repair merely constitutes a breach of contract.

As was said at the beginning of this section on legal aspects, the concept of liability is based upon rights, wrongs and duties inherent in civil conduct. This has been but a brief summary of the many rules of law applicable to such civil conduct, and has been presented to emphasize some of the various types of legal liability which can arise through performance of a construction contract.

SECTION TWO

PLANNING AND ESTIMATING

CHAPTER 10
CONSTRUCTION ESTIMATING

Estimating in construction work may be defined as forecasting the cost of a structure or other project. It is usually accomplished by predicting the amounts of materials, manpower, and equipment that will be required, and multiplying each by its current cost. The time from start to finish of a job must be calculated to obtain total costs.

Estimating is an essential part of all construction work. It is necessary in bidding for contracts, in scheduling of jobs, in planning a project, and in operating the construction company at a profit. Poor or careless estimating is a common cause for business failure in the construction industry.

Furnishing all the knowledge and experience necessary to become a professional estimator is not the purpose of this section. Rather, the intent is to give the overall philosophy of estimating and a general knowledge of the principles involved.

There are a great many methods of performing estimates, and for each method there are a variety of ways of handling the details. The exact method and details of the estimating procedure depend on the preferences of the estimator and the purpose of his estimate. If you are working for someone, you will adapt his procedures, and if you are self-employed, you will select your own methods. For this reason, the methods illustrated in this section need not be memorized, they are only intended to help you get the feel of estimating. When you get on an actual job, you will undoubtedly be given a chance to work with an experienced estimator. Here you will pick up the practical knowledge you need before you go on your own.

The key to successful contracting is carefully estimating cost and bidding only jobs on which you can make a reasonable profit. It is important that your estimate be accurate. If you submit a high bid, you will not get the job. If your estimate is low, you are likely to spend more money than the owner will pay for doing the job. The time estimate is also important, since most contracts require that the job be completed at a certain time. The man-hours, equipment supplies, and materials must be estimated accurately to insure that you have an adequate supply of each to complete the job on schedule without sacrificing your performance on other jobs to which you are committed. These points, eloquently summarized by Frank R. Walker as early as 1915 in the introduction to his *Building Estimators Reference Book,** are equally applicable today. Although many changes and improvements in materials, tools, and equipment have been made since 1915, all the variables Mr. Walker mentions still exist and must be considered carefully in an accurate estimate.

> "The building contractor and estimator engaged in computing the costs of our modern buildings must eliminate the item of guesswork from the estimates as far as possible. This is not an easy matter, as building construction methods can hardly be reduced to the scientific basis of factories and shops, with their fixed surroundings and control of weather vagaries. When the many unforeseen conditions entering into the cost of modern buildings are taken into consideration, that is readily seen. Weather, rain, snow, cold, etc., are all capable of causing inestimable damage and expense, not including the overhead charges for superintendence, etc., which must be paid regardless of weather conditions. A concrete mixer, hoisting engine, derrick or some other piece of equipment may break, causing delays, overtime, or other unforeseen expenses. These are all items over which the contractor or estimator has little control, and he must, therefore, use his best judgment as to costs and methods of construction.
>
> "To the casual observer, or one not understanding building conditions, the prices may appear indefinite or with too great a variation of labor quantities and costs. This is easily explained. A brick mason will lay more brick on a bright warm day than when the thermometer is at freezing.

* Frank R. Walker, *Building Estimators Reference Book,* Frank R. Walker Co., Chicago, 1915.

A carpenter will lay more hardwood flooring in one large room than in ten small rooms requiring the same amount of flooring, but with five to ten times the amount of cutting; a marble setter will set more base in a long straight corridor than in one cut up with many pilasters and with from three to five times the number of pieces containing the same number of lineal feet of base.

"These conditions can be multiplied indefinitely. The estimator should study his plans carefully and base prices accordingly.

"The superintendence of the job will have much to do with the final cost of the work. No two executives employ the same methods of construction or of handling men and few individuals produce the same quantity of finished work. When business is dull good men are easily obtained, but during periods of building prosperity the superintendent has a smaller range of choice and must exercise his best judgment in the selection of men.

"For these reasons the labor cost of any piece of work is a variable item and all future estimates can only be based on past experiences."

Weather conditions, lumber prices, union contracts, characteristics of the building site, labor productivity, temporary structures that may be necessary, financing cost, and the cost of making job investigations and estimates are a few examples of these that existed then and still exist today. Estimates, as you know, are used for a variety of purposes in the construction industry. For this reason, and because unneeded detail in the estimates wastes preparation time and money, there are a number of types of estimates designed for particular uses.

1. *A detailed estimate.* This includes quantities and costs of everything necessary for completion of the job. A detailed estimate may be made by the unit quantity or total quantity method. In the unit quantity method the cost of a part of unit of the job is calculated, then the cost for the complete job is found by multiplying the cost of the unit by the number of units. On a simple repetitive concrete job, for instance, the total cost of one cubic yard of concrete would be found. This unit cost would then be multiplied by the number of cubic yards in the job. This method has the advantage of permitting unit costs on various jobs to be easily compared. The total estimate can also be easily corrected when the size of the total job varies.

   The total quantity method of preparing a detail estimate requires that the amount of each kind of labor material and supplies be estimated. Then these amounts are multiplied by their respective costs, and all these quantities are added up to give the total job cost. On the simple concrete job, for instance, the cost of all the materials, all the labor, and all the equipment would be found separately and added to give the total estimate.

2. *The short-cut or approximate estimate.* This is a rough estimate usually used as a check against a gross error in the detailed estimate. The usual approach is the square foot or square yard and cubic foot or cubic yard procedure. In this method the cost for a square foot or cubic foot of a given type of structure is estimated, then this figure is multiplied by the number of square or cubic feet in the proposed structure. This method is not very reliable and is useful only as a rough check.
3. *A complete estimate.* This includes all costs relating to the work. In addition to the main contract and subcontracts; land, legal fees, architect fees, financing, interest, insurance, and taxes during construction should be added.
4. *A quantity survey.* This is an estimate of the quantities of all materials needed to complete the job. This is sometimes submitted to all contractors that are bidding on a given job. Each then prepares his own estimate of the cost.
5. *A contractor's estimate.* This is a detailed and carefully prepared complete estimate including overhead and profit used for determining the price the contractor will bid.
6. *Progress estimate.* Progress estimates are usually made once a month to determine the current state of the job and are used as a basis for periodic payment to the contractor.
7. *A final estimate.* This is made by the engineer or architect at the completion of the contract to determine the exact quantities of material and installed equipment that have been furnished. This type of estimate is often made before final payment to the contractor. It is a verification that the work has been furnished in accordance with the plans and specifications.

Two categories, costs and time, help simplify the broad field of estimating. Cost estimating will include materials, tools and supplies, and manpower and equipment. The subject of time estimating will include practice in scheduling of work.

Estimating for a small building and for street construction will be included in the text.

COST ESTIMATING

A contractor or estimator will have an opportunity to submit bids on any type of structure. These structures would include roads, streets, bridges, small buildings, apartments or factories, dams, irrigation projects, and many others. To minimize financial risk, the firm would probably bid only on structures or work with which they were familiar, or bid only on portions of projects as a subcontractor, such as foundations for apartment buildings, paving residential streets, or framing of residential buildings.

Since the profit margin on most construction jobs is relatively small, overlooking the smallest detail can mean the difference between profit and loss. Before submitting a bid on a job, it is important to consider every detail on the plans and specifications, the conditions that will be encountered on the building site, and certain needs and wants of your client. The plans for a structure show the details of its construction (Figs. 10-1, 10-2, and 10-3). They include illustrations showing the foundation, the floor layout, the elevations or side views, certain details of specific parts of the building such as stairways, arrangement of the building on the plot, and details of the plumbing and electrical services.

The foundation plan must be carefully examined, especially by the concrete contractor, because it shows the footing size and location, the foundation size and volume, the piers or retaining walls necessary, the footing and drainage details, post and beam locations and size, floor joist location and size, and details of basement windows, partitions, plumbing, heating and electrical equipment. All of this information must be observed in preparing the estimate.

The floor plan gives details on the exterior and interior walls, doors, windows, ceiling framing, plumbing, heating, and electrical equipment. This floor plan is the most detailed of the views; all of its information must be picked up and carefully checked by the estimator. The elevations or side views give the heights, measurements, and information on siding, roofing, and exterior finish. Detail views are used to explain special characteristics of the structure such as sections of walls, corner details, and kitchen, stair, and fireplace construction.

The plot plan shows the overall picture of the building in relation to the land and the utilities that must become a part of the finished project. It illustrates the placement of various structures on the plot, exterior flat work, site improvements, landscaping, existing and finish contours, and drainage. Utilities' distances and connections are also usually shown, although they do not appear in our example.

The specifications are written instructions to the contractor that supplement the plans. They contain important information about the materials, style, workmanship, and finish. The specifications may be written in technical or legal language and they may be one page or several hundred pages depending on the size of the project. The specifications for the small home shown in Figures 10-1, 10-2, and 10-3 are titled "description of materials," and they include FHA dates as noted on their first page. If there is a difference between the information in the plans and the specifications, the contract documents will usually state that the data in the specifications takes precedence.

The contractor should develop a close working relationship with his client or his authorized representative, who may be an architect or an engineer, if he wants harmony and cooperation on the job and his business to grow. Because of the dealings with former customers and clients, he should feel free to discuss his costs openly with the client or his representative. Whenever there is an opportunity to decide between various building methods or materials, the contractor should give the client or his agent an opportunity to make these decisions.

The contractor should also carefully examine the site, noting the soil

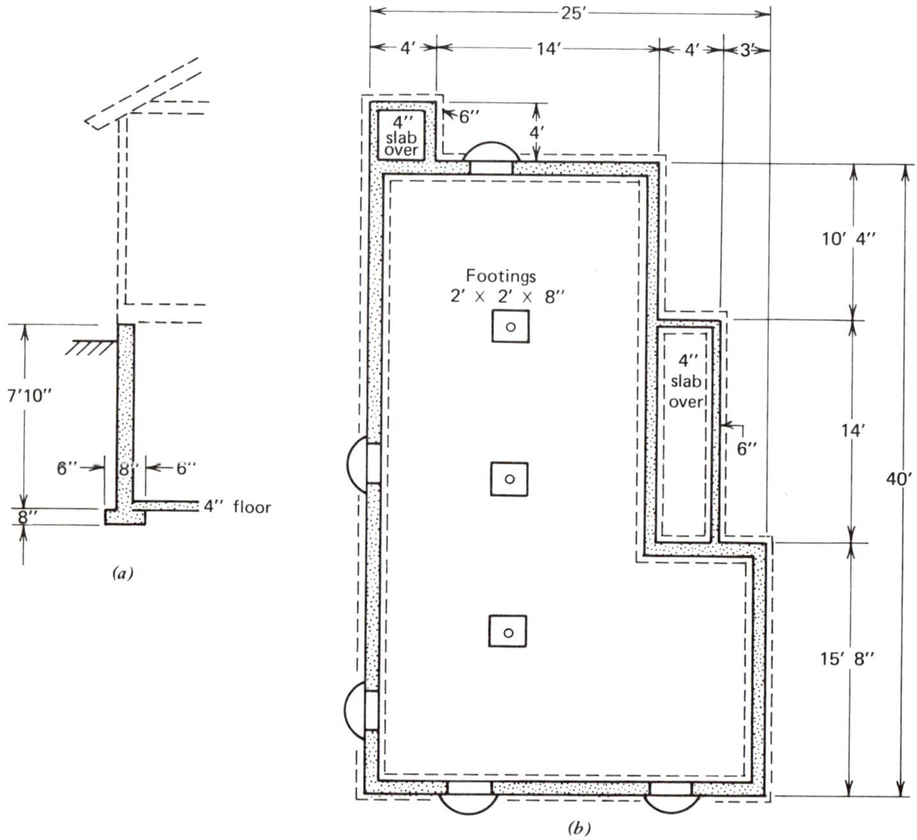

Figure 10–1 (*A*) Wall section. (*B*) Foundation plan.

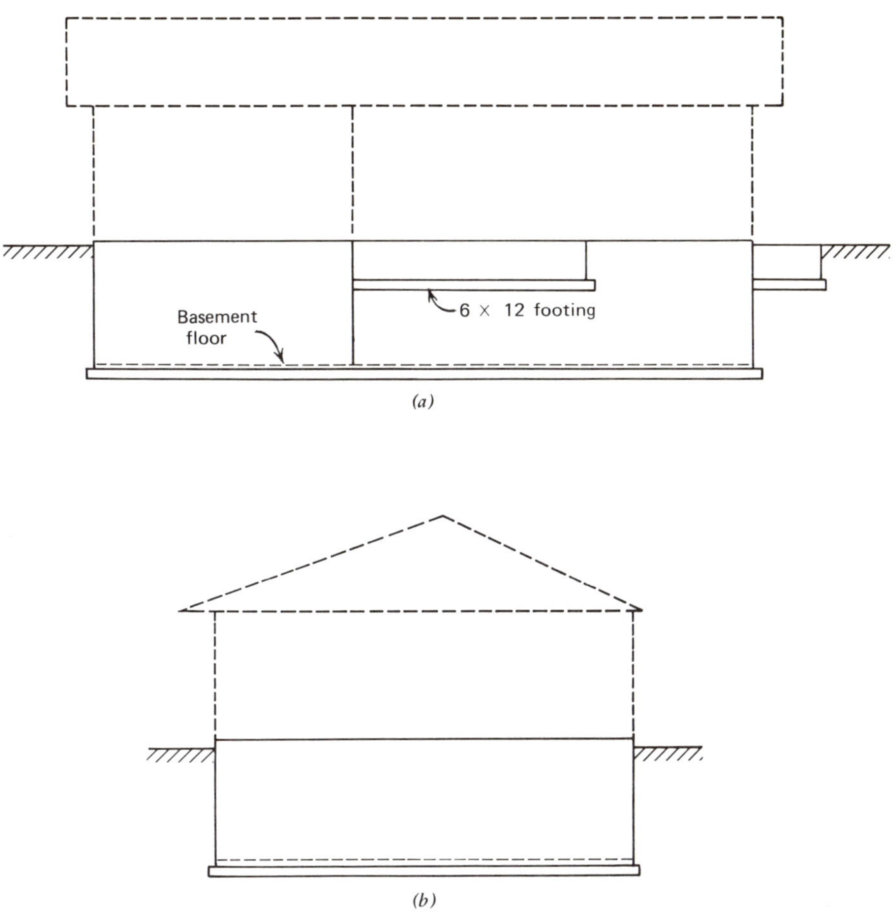

Figure 10–2 (*A*) Front elevation. (*B*) West elevation.

conditions, water table, drainage characteristics, relationship to other buildings, access to the site, natural surroundings, and other physical characteristics.

Weather must also be considered by the contractor. If the job is in the north, he must allow for additional winter job cost and production breaks, which he does not have during the spring and summer months. He must include the cost for heating and protecting winter construction work and attempt to schedule the job so that the house will be closed in to permit inside work during the cold months. It is a well known fact that a workman is more productive in 80°F weather than he is in 10°F weather

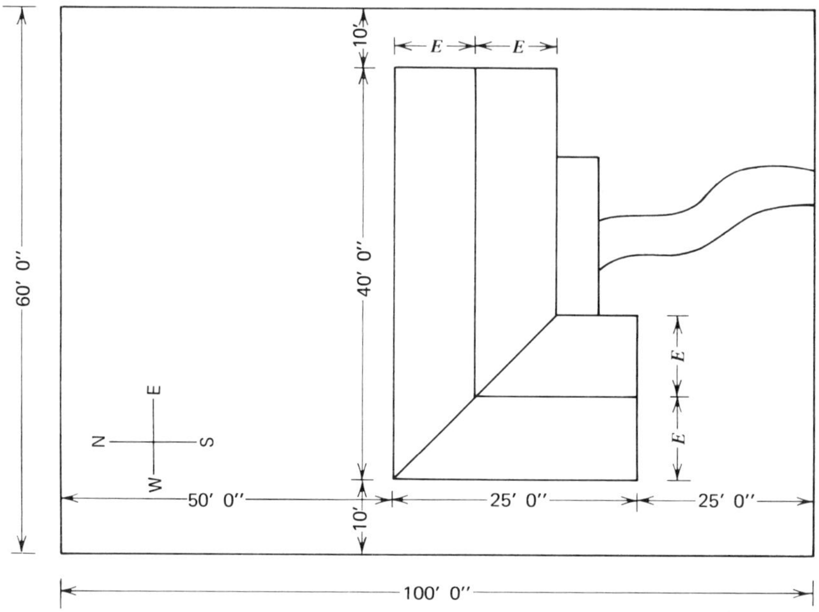

Figure 10–3

when he spends a lot of his time warming himself and when he is handicapped by heavy clothing.

Another important factor to consider in making the estimate is the possibility of fluctuations in costs for labor and materials. If the increase is a certainty, it can be included in the estimate. Although no one can be up to date on every possible fluctuation in price of each part that goes into a structure, protection against losses due to changes in basic contract rates for labor should be included in the estimate. An appropriate strike clause should also be written into the contract so that the builder will be able to receive an allowance for time lost for work stoppages or strikes. Variations in employee efficiency must also be considered. Does the crew you are hiring produce high quality work efficiently or are they sloppy and slow. Are the men well-trained or are there many poorly trained apprentices on the job. These factors can create large discrepancies in both time and cost estimates.

Finally, it is never profitable to rush the calculation of an estimate. More builders have submitted a poor estimate that has ultimately cost them jobs, profits, or losses because they couldn't or didn't take the time to do a good, accurate estimating job. It is important to take the time to check the prices of all materials and supplies, get reliable bids from subcontractors who can furnish payment and performance bonds, and have extensions and calculations double checked by a second person to insure accuracy before you submit a bid for a job. These hours spent on reviewing and checking prices and quotations are hours and dollars well spent.

In addition to estimating the direct costs of the actual work, a number of items must be considered. These include overhead cost, allowance for profit, cost of purchasing and improving the land, and cost of planning the job (Fig. 10-4).

GENERAL OVERHEAD

The items included in general overhead expense have often been neglected by the type of builder who feels that the keeping of detailed records on such expenses are not important. These overhead expenses are vital and should be accounted for in every estimate as well as in the builder's accounting system.

Every builder has certain fixed costs, such as office rent, staff salaries, and telephone expense, which must be paid every month regardless of the amount of work to be done. These expenses can never be directly charged to any particular job, rather they are the basic costs of doing business.

In addition to the static fixed costs that the builder must meet, there are other variable costs that cannot, without a great deal of time and effort, be attributed to any particular job. It is often more expedient and convenient to treat these as part of overhead costs rather than trying to allocate specific costs to each and every individual job.

A partial list of the typical overhead expenses that might be incurred by the builder in fixed and variable costs would include the following items: office rent; fuel and light; telephone; advertising; donations and charity; professional periodicals; professional association fees; legal services; business consultants; office salaries; executive salaries; office supplies; automobile expense; travel and entertainment expense; business insurance; business taxes; licenses and operating fees; maintenance; workmen's compensation insurance and unemployment compensation for main office personnel; and general liability insurance.

If annual gross sales average \$150,000 and overhead \$12,000, the builder should figure that every estimate should carry overhead of 8% to cover his overhead costs. Should he find that his gross sales for the current year were more than gross sales for the previous year, then he might reduce this percentage; however, he should start on the basis of some financial budget that would give him a sound budgeting procedure to follow.

JOB OVERHEAD

The costs of job overhead are often overlooked in the estimating process. However, the costs of planning a job or project are just as important, for the sake of accuracy, as the costs of ceiling joists. The builder often likes to lump all of these costs into one neat package—job overhead. But they can—and should—be estimated individually and separately, whenever possible. The job overhead costs divide into the following major sections: building permits and fees; job insurance and barricades and watchman service; liability protection; contractor's bonds; construction heat, water and electricity; engineering fees; temporary buildings and sheds.

Within each of these major categories are numerous costs that are integral parts of the job costs and should be estimated individually, in order that the builder can get a realistic picture of the costs that he may expect to incur during the construction of the house. Little is to be gained by trying to lump all of these costs into one unit such as overhead, for each job will require special attention to some of these particular details.

Field Design

There are various costs that the builder incurs, whether he does the field design or has the work done. If the builder does the work himself or has his own organization do it, he must meet the costs of drafting labor, drafting materials, and preliminary survey costs. In addition to these costs, he may incur additional costs for consulting fees from engineers or architects who prepared the project plans.

Building Permits and Fees

The cost of obtaining the building permits from local and state governmental agencies is a part of the job costs. These permit and fee costs can be determined easily prior to the work, and it is essential that they be considered in the estimate, unless the owner is to assume the costs for these fees.

The costs of the building permits and fees are usually based upon the extent and cost of the work to be performed and will vary from city to city in cost and the number of permits that must be obtained. Some cities charge a flat fee per house; while others base their fee upon the cost of

☐ Proposed Construction

☐ Under Construction

DESCRIPTION OF MATERIALS

No.______________
(To be inserted by FHA)

Property address ______________________ City__________ State__________

Mortgagor or Sponsor ____
(Name) (Address)

Contractor or Builder ____
(Name) (Address)

The specifications include this Description of Materials, the applicable FHA Minimum Construction Requirements, and any applicable FHA Engineering Bulletin

INSTRUCTIONS

1. Describe materials and equipment to be used by entering the essential information and marking an X in each pertinent check-box.
2. Do not include alternates or "or equal" items. Acceptance of substitute materials or equipment can be requested of FHA, through the mortgagee, before or after construction has begun.
3. Unless otherwise described or shown, required work will be assumed to be the minimum acceptable to FHA.
4. Construction which exceeds FHA minimum requirements will be recognized in the FHA cost estimate only if adequately described herein.
5. Signature of mortgagor or sponsor is required on page 4.

1. EXCAVATION:
   Bearing soil, type ____

2. FOUNDATIONS:
   Footings: Concrete mix ______________ Reinforcing ______________
   Foundation wall: Material ______________ Reinforcing ______________
   Interior foundation wall: Material ______________ Party foundation wall ______________
   Columns: Material and size ______________ Piers: Material and reinforcing ______________
   Girders: Material and sizes ______________ Sills: Material ______________

DESCRIPTION OF MATERIALS (cont.)

Basement entrance areaway ______ Window areaways ______
Waterproofing ______ Footing drains ______
Termite protection ______
Basementless space: Ground cover ______ Insulation ______ Foundation vents ______
Special foundations ______
______
______

3. CHIMNEYS:
Material ______ Prefabricated (*make and size*) ______
Flue lining: Material ______ Heater flue size ______ Fireplace flue size ______
Vents (*material and size*): Gas or oil heater ______ Water heater ______
______

4. FIREPLACES:
Type: ☐ Solid fuel; ☐ gas-burning; ☐ circulator (*make and size*) ______ Ash dump and clean-out ______
Fireplace: Facing ______; lining ______; hearth ______; mantel ______
______

5. EXTERIOR WALLS:
Wood frame: Grade and species ______ ☒ Corner bracing. Building paper or felt ______
Sheathing ______; thickness ______; width ______ ☒ solid; ☐spaced ______" o.c.; ☐ diagonal; ______
Siding ______; grade ______; type ______; size ______; exposure ______"; fastening ______
Shingles ______; grade ______; type ______; size ______; exposure ______"; fastening ______
Stucco ______; thickness ______". Lath ______; weight ______lb.
Masonry veneer ______ Sills ______ Lintels ______
Masonry: Facing ______; backup ______ thickness ______". Bonding ______
Door sills ______ Window sills ______ Lintels ______
Interior surfaces: Dampproofing, ______ coats of ______; furring ______
Exterior painting: Material ______; number of coats ______
Gable wall construction: ☐ Same as main walls; ☒ other ______
______

6. FLOOR FRAMING:
Joists: Wood, grade and species ______; other ______; bridging ______; anchors ______
Concrete slab: ☒ Basement floor; ☐ first floor; ☐ ground supported; ☐ self-supporting; mix ______; thickness______";

reinforcing ____________; insulation ____________; membrane ____________
Fill under slab: Material ____________; thickness ____________". ____________

7. SUBFLOORING: (*Describe underflooring for special floors under item 21.*)
Material: Grade and species ____________; size ____________; type ____________
Laid: ☒ First floor; ☐ second floor; ☐ attic ____________ sq. ft.; ☒ diagonal; ☐ right angles. ____________

8. FINISH FLOORING: (*Wood only. Describe other finish flooring under item 21.*)

Location	Rooms	Grade	Species	Thickness	Width	Bldg. Paper	Finish
First floor							
Second floor							
Attic floor ____ sq. ft.							

9. PARTITION FRAMING:
Studs: Wood, grade and species ____________ Size and spacing ____________ Other ____________

10. CEILING FRAMING:
Joists: Wood, grade and species ____________ Other ____________ Bridging ____________

11. ROOF FRAMING:
Rafters: Wood, grade and species ____________ Roof trusses (see detail): Grade and species ____________

12. ROOFING:
Sheathing: Grade and species ____________; size ________ type ________; ☒ solid; ☐ space ________" o.c.
Roofing ____________; grade ________; weight or thickness ________; size ________; fastening ________
Stain or paint ____________ Underlay ____________
Built-up roofing ____________; number of plies ________; surfacing material ________
Flashing: Material ____________; gage or weight ________; ☐ gravel stops; ☐ snow guards

13. GUTTERS AND DOWNSPOUTS
Gutters: Material ____________; gage or weight ________; size ________; shape ________
Downspouts: Material ____________; gage or weight ________; size ________; shape ________; number ________
Downspouts connected to: ☐ Storm sewer; ☐ sanitary sewer; ☐ dry well. ☒ Splash blocks: Material and size ________

DESCRIPTION OF MATERIALS (cont.)

14. LATH AND PLASTER
Lath ☐ walls, ☐ ceilings: Material ________; weight or thickness ________ Plaster: Coats ________; finish ________
Dry-wall ☒ walls, ☒ ceilings: Material ________; thickness ________; finish ________; joint treatment ________

15. DECORATING: (*Paint, wallpaper, etc.*)

Rooms	Wall Finish and Application	Ceiling Finish Material and Application
Kitchen		
Bath		

16. INTERIOR DOORS AND TRIM:
Doors: Type ________; material ________; thickness ________
Door trim: Type ________; material ________ Base: Type ________; material ________; size ________
Finish: Doors ________; trim ________
Other trim (*item, type and location*) ________

17. WINDOWS:
Windows: Type ________; make ________; material ________; sash thickness ________
Glass: Grade ________; ☐ sash weights; ☐ balances, type ________; head flashing ________
Trim: Type ________; material ________ Paint ________; number coats ________
Weatherstripping: Type ________; material ________ Storm sash, number ________
Screens: ☐ Full; ☒ half; type ________: number ________; screen cloth material ________
Basement windows: Type ________; material ________; ☒ screens, number ________; ☐ Storm sash, number ________
Special windows ________

18. ENTRANCES AND EXTERIOR DETAIL:
Main entrance door: Material ________; width ________; thickness ________ Frame: Material ________; thickness ________
Other entrance doors: Material ________; width ________; thickness ________ Frame: Material ________; thickness ________
Head flashing ________ Weatherstrippnig: Type ________; saddles ________
Screen doors: Thickness ________; number ________; screen cloth material ________ Storm doors: Thickness ________"; number ________

Combination storm and screen doors: Thickness ______"; number ______; screen cloth material ______
Shutters: ☐ Hinged; ☐ fixed. Railings ______ Louvers ______
Exterior millwork: Grade and species ______ Paint ______; number coats ______

19. CABINETS AND INTERIOR DETAIL:
Kitchen cabinets, wall units: Material ______; lineal feet of shelves ______; shelf width ______
Base units: Material ______; counter top ______; edging ______
Back and end splash ______ Finish of cabinets ______; number coats ______
Medicine cabinets: Make ______; model ______
Other cabinets and built-in furniture ______

20. STAIRS:

Stair	Treads		Risers		Strings		Handrail		Balusters	
	Material	Thickness	Material	Thickness	Material	Size	Material	Size	Material	Size
Basement										
Main										
Attic										

Disappearing: Make and model number ______

21. SPECIAL FLOORS AND WAINSCOT:

Floors	Location	Material, Color, Border, Sizes, Gage, Etc.	Threshold	Base	Underfloor
	Kitchen				
	Bath				

Wainscot	Location	Material, Color, Border, Cap, Sizes, Gage, Etc.	Height	Height at Tub	Height at Shower
	Bath				

Bathroom accessories: ☒ Recessed; material ______; number ______; ☒ attached; material ______; number ______

DESCRIPTION OF MATERIALS (cont.)

22. PLUMBING:

Fixture	Number	Location	Make	Mfgr's Fixture Identification No.	Size	Color
Sink						
Lavatory						
Water closet						
Bath tub						
Shower over tub*						
Stall shower**						
Laundry trays						

*☐ Curtain rod **☐ Door ☐ Curtain rod

Water supply: ☒ Public; ☐ community system; ☐ individual (private) system.

Sewage supply: ☒ Public: ☐ community system; ☐ individual (private) system.

House drain (inside): ☒ Cast iron; ☐ tile; ☐ other ______ House sewer (outside): ☐ Cast iron; ☐ tile; ☐ other ______

Water piping: ☐ Galvanized steel; ☒ copper tubing; ☐ other ______ Sill cocks, number ______

Domestic water heater: Type ______; make and model ______

recovery ______ gph. 100° rise. Storage tank: Material ______; capacity ______ gallons.

Gas service: ☒ Utility company; ☐ liq. pet. gas; ☐ other ______ Gas piping: ☒ Cooking; ☒ house heating.

Footing drains connected to: ☐ Storm sewer; ☐ sanitary sewer; ☐ dry well. Sump pump ______

______

23. HEATING:

☐ Hot water. ☐ Steam. ☐ Vapor. ☐ One-pipe system. ☐ Two-pipe system.

☐ Radiators. ☐ Convectors. ☐ Baseboard radiation. Make and model ______

Radiant panel: ☐ Floor; ☐ wall; ☐ ceiling. Panel coil: Material ______

☐ Circulator. ☐ Return pump. Make and model ______; capacity ______ gpm.

Boiler: Make and model ______ Output ______ Btuh.; net rating ______ Btuh.

______

Warm air: ☐ Gravity. ☒ Forced. Type of system ______

Duct material: Supply ______; return ______ Insulation ______, thickness ______ ☐ Outside air intake.

Furnace: Make and model ______ Input ______ Btuh.; output ______ Btuh.

☐ Space heater; ☐ floor furnace; ☐ wall heater. Input ________ Btuh.; output ________ Btuh.; number units ______
Make, model ____

Controls: Make and types ____

Fuel: ☐ Coal; ☐ oil; ☐ gas; ☐ liq. pet. gas; ☐ electric; ☐ other ____________; storage capacity ____________

Firing equipment furnished separately: ☐ Gas burner, conversion type. Stoker: ☐ Hopper feed; ☐ bin feed.
Oil burner: ☐ Pressure atomizing; ☐ vaporizing ____________________
Make and model ____________________ Control ____________________

Electric heating system: Type ____________ Input ______ watts; @ ______ volts; output ______ Btuh.

Ventilating equipment: Attic fan, make and model ____________________; capacity ________ cfm.
Kitchen exhaust fan, make and model ____________________
Other heating, ventilating, or cooling equipment ____________________

24. ELECTRIC WIRING
Service: ☒ Overhead; ☐ underground. Panel: ☐ Fuse box; ☒ circuit-breaker ____________ Number circuits ______
Wiring: ☐ Conduit; ☐ armored cable; ☒ nonmetallic cable; ☐ knob and tube; ☐ other ____________
Special outlets: ☐ Range; ☐ water heater; ☐ other ____________________
☐ Door bell. ☒ Chimes. Push-button locations ____________________

25. LIGHTING FIXTURES:
Total number of fixtures ____________ Total allowance for fixtures, typical installation, $____________
Nontypical installation ____

26. INSULATION:

Location	Thickness	Material ,Type and Method of Installation	Vapor Barriers
Roof			
Ceiling			
Wall			
Floor			

DESCRIPTION OF MATERIALS (cont.)

27. MISCELLANEOUS:

(*Describe any main dwelling materials, equipment, or construction items not shown elsewhere*): ______

HARDWARE: (*Make, material, and finish*) ______

SPECIAL EQUIPMENT: (*State material or make and model*)

Venetian blinds ______ Number ______ Automatic washer ______
Kitchen range ______ Clothes drier ______
Refrigerator ______ Other ______
Dishwasher ______
Garbage disposal unit ______

PORCHES:

TERRACES:

GARAGES:

____

____

____

____

____

WALKS AND DRIVEWAYS:

Driveway: Width ______ Base material ________; thickness ____". Surfacing material ______________; thickness ____"

Front walk: Width _____ Material __________; thickness ____". Service walk: Width ____ Material _____; thickness ____"

Steps: Material ______________________; treads ______"; risers ______". Check walls __________

____

OTHER ONSITE IMPROVEMENTS:

(*Specify all exterior onsite improvements not described elsewhere, including items such as unusual grading, drainage structures, retaining walls, fence, railings, and accessory structures.*)

____

____

____

____

____

LANDSCAPING, PLANTING, AND FINISH GRADING:

Topsoil ______" thick: ☐ Front yard; ☐ side yards; ☐ rear yard to ________ feet behind main building.

Lawns: (*seeded, sodded, or sprigged*) ☐ Front yard __________; ☐ side yards __________; ☐ rear yard __________

Planting: ☐ as specified and shown on drawings; ☐ as follows:

______ Shade trees, deciduous, ______" caliper.

______ Low flowering trees, deciduous, ______' to ______'

______ High-growing shrubs, deciduous, ______' to ______'

______ Medium-growing shrubs, deciduous, ______' to ______'

______ Low-growing shrubs, deciduous, ______' to ______'

______ Evergreen trees, ______' to ______', B & B.

______ Evergree shrubs, ______' to ______', B & B.

______ Vines, 2-year __________

____

____

IDENTIFICATION: *This exhibit and all others submitted with the same application must be identified by the signature of the applicant* mortgagor *if firm commitment is applied for, or of the* sponsor *if conditional commitment is applied for.*

Date ______________, 19___ Signature ______________________________

☐ *Mortgagor* ☐ *Sponsor*

construction; and still others upon the number of fixtures and outlets that go into the house. Regardless of the type and amount of fees, the builder should ascertain the costs and include them in the estimate.

Personnel

There may be a number of men on the job site whose labor is not direct job cost but rather job overhead. Their responsibilities may be job superintendent, timekeeper, field office man, or mechanic. The number of men needed will depend on the size of the job. On small projects, one man may handle all these tasks.

Finance Charges

In those cases where the project builder and the custom builder obtain the permanent and temporary financing for the client, the builder must be cognizant of the cost of financing. Under the current system of discounting permanent financing and because of the high cost of construction loan money, this aspect can assume a major dollar role in the cost of construction and is not a charge that the builder can afford to overlook or forget. There are three basic methods of real estate finance: conventional financing channels, FHA insurance, and VA guarantee. The charges in the three types of loans vary because of legislation, business practice, and geographic area.

Job Insurance and Liability Protection

The builder who assumes the responsibility of keeping insurance on the buildings under construction must have builder's risk insurance to give complete protection. Such items as fire, extended coverage, vandalism, and malicious mischief are included. These costs are necessary, unless the builder wishes to gamble or risk his profits, business capital, and personal finances. For this reason, the builder should carry and include in the estimate the costs of these insurance policies.

Contractor's Bonds

A builder who is bidding a job will be required to furnish bid bonds, if stipulated by bidding documents, and maintain completion or performance bonds during construction. If he is the successful bidder and enters into a construction contract, he will be required to furnish performance and payment bonds.

Although these bonds are more common in industrial or commercial construction, the residential builder may find these bonds required by architects and by the client's attorney. The cost of such bonding documents may be as much as 1% of the cost of the completed structure. Thus, for a $30,000 residence, the bond might cost $300. There is no small item in the estimate and failure to include it would mean that if the remainder of the estimate were correct, you would sacrifice $300 of the profit to which you were entitled.

Construction Heat, Water, Electricity, and Telephone

In order to operate power equipment and have construction water and job communications, it will be necessary that you have water, electricity, and telephone on the job; and in the event of winter work, you will need construction heat. Failure to include these items will not bankrupt a company, but they reduce profits; so the same care and consideration in estimating these items must be taken as on all other incidentals.

Engineering Fees and Testing

If the job calls for engineering and testing work either before, during, or after construction, the costs of the engineering or testing services and work should be considered in the estimate. An example of this type of expense would be a survey required by the architect, to be made directly after the foundation had been placed. Another example might be test cylinders for concrete work.

Temporary Buildings and Sheds

On some jobs, it may be necessary to put up material, field office, and convenience sheds, especially where inclement weather otherwise would appreciably lower the efficiency of the crew. If so, then the construction costs of these buildings, less their resale or salvage value, should be included in the estimate, as these costs are part of your job expenses.

Job overhead costs may be sizable for a project builder and for the custom builder who takes on many of the responsibilities for design and preparation of the residential construction. Not all of these items will appear in every estimate for every building, but be certain that you have at least reviewed these major topic areas for the possibility of including them in the estimate.

PROFIT

This is one of the most challenging areas in the field of estimating. The builder who asks what his profit percentage should be, must first take a good look at his organization and investment to see what his basic policy is in regard to this matter. For the builder who anticipates building a large volume of housing and meeting a low-income market, his percentage of profit may be considerably lower than the custom builder who anticipates building only a few, and who has a good reputation for quality construction. His workload will also influence him; he will be willing to accept less profit in slack periods than in busy ones. The market itself will have much to say about the amount of profit that the builder can make and, if he is working on competitive bidding, his competition will tend to set a maximum on profits that he cannot exceed if he is going to compete.

CHAPTER 11
ESTIMATING

USE OF THE ESTIMATE

The estimate should be made an integral part of the office records. It should not be a one-time bidding guide, but should, if properly prepared, be used throughout the entire project operation. In addition to giving the contractor a sound basis for bidding and building, the estimate should provide the following aids in other phases of business:

1. *Purchasing.* From the estimate the builder should be able to effectively purchase his materials without any new takeoff being required.
2. *Scheduling.* The builder should use the estimate to help schedule delivery of materials, crews, and crew size of both his own labor and that of his subcontractors.
3. *Supervision.* From the estimating data and cost records of past jobs, the builder can increase the efficiency of his crews and supervision, since with these two records he has available the knowledge of past performance on costs and production.
4. *Accounting.* The builder's estimating system should be coordinated in every aspect with that of his accounting system, from daily time and production reports and check stubs to the final cost summary.
5. *Finance.* With most of the phases of construction coordinated to the estimating system, the builder is able to forecast his financial needs,

both for short-term and possible long-term financing, in advance of any crisis. This allows the business to flow along smoothly and to make the most effective use of working capital and lending policies.

SITE DEVELOPMENT

When site development begins, people passing by the land will be able to see that the job is under way. The site must be prepared for the structure that is to be built upon it.

There are nine major categories included in the consideration of land costs: (1) Land zoning, (2) surveying and legalities, (3) engineering or site surveying, (4) water systems, (5) sanitary sewer systems, (6) gas systems, (7) electrical systems, (8) street and walk improvements, (9) storm sewer. Many of these items require professional services by individuals or companies such as attorneys, engineers or architects, and are normally done on a professional contract basis. These utilities and services are normally subcontracted. The costs charged to the job resulting from these subcontracts include not only materials and labor, but also the profit and overhead figures of the subcontractors.

Land is unique: it is immobile and indestructible, and topographic features and soil conditions vary greatly from community to community. Because of these characteristics, the estimator who strives for accuracy and completeness must carefully consider many variables to insure that the site will be suitable.

Certain utilities and site improvements must be put in before the land is ready for residential use. The modern residence must have streets, curbs and gutters, sidewalks, sanitary sewers, water, gas, and electricity. The cost of bringing these utilities to the site and to the house is often more expensive than the land itself. Therefore, careful estimating techniques must be undertaken if the builder is to start his estimate correctly.

Before the estimate is begun, the site must be carefully examined to determine whether or not it is suitable for the required construction. Also it should be examined to see if there are any special features of the site that must be given special consideration in the estimate. Many localities have adverse topography or soil conditions or other problems that can only be corrected through expensive grading, structural designing, drainage and solidification processes. Be sure that your examination of the site is complete and where there are doubts, check with qualified individuals.

In making the estimate, there are several documents that will be helpful; one of these is a checklist. This is a list of every item, part, and test necessary to complete a certain type of job, such as building a house, bridge, or retaining wall. You will prepare a list that includes almost every possible item you may encounter once you have developed an area of specialization on the job. This checklist serves as a reminder that every item, part, or test is included and insures that a small but often important item will not be overlooked. The old estimating rule that "The builder should include something for every item" is still basically sound. A builder who forgot to estimate the masonry fireplace, or failed to include the finished hardware, or forgot the basement flatwork will agree that the checklist is the cheapest piece of record that he can own. The checklist is also important in following through on purchasing to insure that each of the small items is there when the craftsman needs them. Table 11-A is a check list for site development.

Table 11-A
CHECKLIST FOR SITE DEVELOPMENT

Operation	Materials or Type of Work	Equipment	Labor
Legal and Engineering			
Attorney's title opinion			
Abstracting			
Recording			
Preparation of contracts			
Land			
Raw acreage			
Building site			
Clearing site		Bulldozers	Foreman
	Removing trees	Trucks	Truck drivers
	Removing stumps	Loaders	Machine operators
	Hauling debris	Lifts	Common labor
Wrecking buildings			
	Wrecking	Cranes	Foreman
		Trucks	Operators
	Hauling material from the site	Compressors	Carpenters
		Jackhammers	Common labor
			Truck drivers
		Motorgraders	
Leveling and grading		Bulldozers	Machine operators
		Scrapers	Blaster
		Trucks	
		Loaders	
Real estate closings			
Title insurance			
Engineering			
Surveying			

Table 11-A (cont.)

Operation	Materials or Type of Work	Equipment	Labor
Engineering (continued)			
Subdivision planning			
Soil testings			
Contour detailing			
Utilities			
Water systems			
Public water systems			
	Trenching	Trencher	Machine operators
		Backhoe	Plumbers
	Pipe	Dozer	Common labor
	Valves	Loaders	
	Hydrants	Trucks	
Private well system			
	Well drilling	Drilling Rig	
	Casing		
	Pump		
	Pump building		
	Pressure tank		
	Trenching		
	Pipe		
	Valves		
	Hydrant		
Sanitary sewer			
Public sewer system			
	Trenching		
	Pipe		
	Manholes		
	Lampholes		
	Taps		
Individual sewer system			
	Excavation		
	Trenching		
	Septic tank		
	Sewer tile (Leaching field)		Drain layer
Gas system			
Public gas system			
	Trenching		
	Pipe	Tamper	
	Pressure		
	Taps		
Individual gas system			
	Excavation		
	Trenching		
	Pipe		
	Storage tanks and valves		

Table 11-A (cont.)

Operation	Materials or Type of Work	Equipment	Labor
Electrical systems			
Public electrical			
System overhead	Poles	Drilling machine	Machine operators
	Wire		Common labor
	Insulators		Lineman
	Transformers		Electrician
Underground			
	Trenching	Trencher	
	Cables		
Private electrical system	Generator		
	Motor		
	Wire		
	Fuel storage tank		
Street and walk improvements			
Curb and gutter			
	Excavation	Dozer	Machine operator
	Grading	Loader	Carpenter
	Forming	Truck	Cement finisher
	Concrete	Mixer	Common labor
Street			
	Excavation	Dozer	Machine operator
	Grading	Loaders	Carpenter
	Leveling	Trucks	Cement finisher
	Compacting	Compacters	Common labor
	Forming	Mixer	Foreman
	Fill (bituminous surfacing)		
Sidewalk			
	Excavating	Dozer	Machine operator
	Forming	Loader	
	Fill	Truck	
	Concrete	Mixer	
Storm sewer			
	Trenching	Trencher	Machine operator
	Pipe	Crane	Mason
	Manholes	Truck	Carpenter
	Catch basins	Dozer	Cement finisher
	Head walls		Common labor
			Foreman

CONSTRUCTION

The calculation of the cost of the actual construction can be divided into five parts: materials, supplies, labor, equipment, and subcontracts.

Materials

Materials are items that become part of the structure, such as brick, lumber, concrete, doors, and hardware. In estimating materials, the first step is to list the quantity of each item that will be used on the job. The preparation of this list is called the quantity takeoff. When the takeoff is complete, the cost of each item is entered opposite the quantity. The product of quantity and cost is entered in a third column. These values are total estimated costs of materials for the job.

Supplies

Supplies are items necessary in doing the work that do not become part of the finished job. Examples are form lumber, workmen's boots, work ladder and scaffolds, first aid supplies, form oil, and stakes and chalk line. These items can easily be overlooked, but their costs can be high enough to put a profitable job in the loss column. A checklist should be used to insure that all these items are included in the estimate.

Labor

Estimating labor cost is more than multiplying the average hourly wage of the worker, lead man, or foreman by the number of hours they work. Allowances must be made for variations in wages, changes in working conditions, differences in skill levels, crew size, the experience, interest and teamwork of the workers, the length of the working day, and insurance, payroll taxes, and fringe benefits.

Equipment

Machinery, tools, temporary buildings, and structures must all be considered in estimating equipment or plant cost for a construction job. For a concrete structure, machinery and tools for making forms, bending and cutting reinforcement, storing and handling materials, and mixing, transporting and placing concrete may be necessary. If the equipment is leased or rented, the actual rental or lease cost will be entered. If it is owned, the ownership costs such as the interest on the investment insurance, taxes, permits and fees, and depreciation must be included. The labor cost of machine operators must be included in labor costs. Equipment cost can be based on time or output.

Operating costs such as repairs, fuel oil, and grease must also be added. These figures are normally computed by the owner for each piece of equipment and made available to the estimator.

Subcontracts

When the company that is responsible for doing a job under a prime contract hires another company or professional to do a part of the job, he is subcontracting the work, and the organization doing that work is a subcontractor. The costs of this subcontracted work become a part of the total estimate and must be carefully checked for accuracy and reliability.

In estimating the cost of the construction of a structure, it is best to break the building process down into steps. In the case of a small house these steps would include the foundation, the rough floor construction, exterior walls, interior walls, roof construction, interior finish, paint, tile, and floor coverings, flat work, site improvement, and utilities. It is not the purpose of the book to examine all the details of estimating for a small house, bridge, or any other type of construction, and we shall therefore omit some of these elements. However, since we are primarily interested in concrete in this book, let's go through the details involved in estimating the costs of the foundation. The method developed for this element can be applied to the other construction steps.

FOUNDATION

This estimating system takes the foundation estimate from the layout through the waterproofing. Items are estimated, piece by piece, as they would be constructed. This procedure carefully picks up the minor as well as the major phases of residential construction, allowing the estimator-builder to meet, formulate, and understand the particular problems as he comes to them.

The major sections that must be estimated for the foundation of every residential and commercial structure are (1) layout, (2) excavation, (3) concrete footings and piers, (4) concrete foundation walls, and (5) waterproofing.

Layout

After the survey has been completed and the engineer has "pinned" or staked out the corners of the property, the builder must lay out the structure for placement of the residence or building and the excavation. Because of the possibility of an error that may affect the entire structure and possible repercussions from neighbors, zoning departments, building departments, and financial institutions, this item should not be hurried.

Excavation

The unit used for estimating the excavation is the cubic yard. A cubic yard is 27 cu ft, and the excavation volume is found by multiplying the length, in feet, by the width, in feet, by the depth, in feet, and dividing this product by 27.

An extra 2 ft on each side of the excavation is allowed for working room for the carpenters in forming and for the waterproofers to do their work. This must be included in the cubic content of the excavation and also for the backfill.

One of the major problems will be the determination of the type of equipment that is to be used in excavating the foundation. There are a number of types that the builder may use; however, as in every building operation, there is a time and a place for the right equipment. Equipment to be considered includes front-end loader, trencher, backhoe, shovel, dragline, grader, bulldozer, and some hand tools.

Concrete Footings and Piers

The unit used in estimating for concrete footings and piers is also the cubic yard.

Footings are employed to distribute the load transmitted to them from the foundation or structure over a larger ground area. Their benefit lies in the fact that a blunt, wide object will not pierce the earth as readily as a narrow or pointed object. In other words, the unit pressures or loads are decreased below the load-carrying capacity of the underlying solids by adequate footings. The plans should be checked very carefully for the size and placement of both footings and piers and for solidification and reinforcing details.

The size of the footings may vary according to soil conditions, local building code specifications, and design. To aid in calculating the volume of a perimeter footing, you should multiply the perimeter of the house by the width by the thickness, each measured in feet, and divide the product by 27 to find the cubic yardage.

Footings may be formed, partially formed, or just excavated.

Usually, if the builder has ready mixed concrete available, he will use this source.

Concrete Foundation Walls

In the estimating process for the concrete foundation walls, there are six distinct operations:

1. Forming
2. Placing reinforcing steel
3. Installing doors, window frames, or blockouts
4. Mixing concrete or buying ready mixed
5. Placing concrete
6. Stripping forms

Forming. The measurement unit used throughout this estimating method, and which is most common throughout the industry, is one square foot of form contact area. This means that both sides of the foundation must be calculated in determining the form contact area for a particular foundation. Normally the surface area of one side can be multiplied by two. This method for figuring the corner measurements is not exact, but it is adequate for the purpose of estimating.

There are numerous methods of forming foundation walls; some use the conventional system of board sheathing or plywood, upright studs, sills, braces, stiffeners and stakes, while other builders have either purchased, built, or rented panel systems of plywood and steel. Each method requires different data for preparing estimates. You will use the data that fits your procedure of forming the foundations. The reuse factor (number of time forms may be reused without refabrication) will often determine the form system to be used.

Placing Reinforcing Steel. Because foundation designs have different sizes, shapes, or locations of reinforcing bars, it is very difficult to give a pound per square foot of steel that will be accurate and realistic. For this reason, it is recommended that reinforcing for each foundation be calculated separately and that the unit of measurement for material and placement be converted to pounds. Provision should be made for laps for splices, ties, chairs, spacers, or welding. In many of the large metropolitan areas, there are subcontractors who specialize in the placement of the steel reinforcing. However, for most residential work, the amount used is relatively small and does not warrant the use of subcontractors for this phase of construction.

Installing Doors, Window Frames, or Blockouts. Allow time for placing the framing members or blockouts in the forms, as well as materials for such members as basement doors and basement windows. Deduct from the concrete quantity estimate the volume used by such doors and windows. If the plans involve special features, such as areaways, be sure that proper consideration is given to the time and materials needed.

Allowance should also be made for the time and materials needed in placing additional embedded items, such as conduit, drains, anchor bolts, and foundation bolts, but no volume need be deducted.

Mixing Concrete or Buying Ready Mixed. The unit of measurement for concrete is the cubic yard, which is 27 cubic feet.

Allowances should be made for all losses and waste when estimating concrete quantities. Additional allowances should be made for stockpile losses when job-mixed concrete is used. Loss factors will vary depending on past experience and the nature of the job. These losses could vary from 5 to 25%.

Placing Concrete. The placing of concrete usually involves wheeling. In some cases, however, concrete is placed with cranes and buckets, pumps, conveyors, chutes, or power buggies. The cost of the equipment and labor for the method used must be included in the estimate.

Stripping Forms. In estimating, costs of stripping forms, labor, supplies, equipment, and so on are included, but materials are omitted since none are used.

Waterproofing

Good quality concrete is the best way to achieve a waterproof structure. To supplement this, the waterproofing or dampproofing of concrete foundations and walls is generally accomplished by one of the following techniques:

1. Liquid repellent preparations used as admixtures in the concrete
2. Parging or plastering
3. Membrane
4. Coatings such as asphalt or pitch (applied hot or cold, brushed or sprayed)
5. Paints

Table 11-B is the checklist for construction of the foundation. As a check on yourself, make an entry for every item to insure that nothing is overlooked. If the item does not apply, note this on the checklist.

Table 11-B
CHECKLIST FOR FOUNDATION CONSTRUCTION

Operation	Supplies and Materials	Equipment (regular hand and power tools not listed)	Type of labor
Layout	Stakes Batterboards Line Nails	Transit or level	Engineer Foreman Carpenter Common labor
Excavation		Bulldozer Trencher Backhoe Power shovel Front-end loader	Foreman Common labor Machine operators
Shoring	Timbers Boards Nails		Foreman Carpenter Common labor
Footings Forming and stripping	Form panels Dimension boards Stakes Form oil Nails		Foreman Carpenter Common labor
Stabilization Solidification	Crushed rock Chemical solution, sand		Foreman Common labor
Reinforcing	Reinforcing steel Steel pins Chairs, ties, and spacers		Foreman Steelworkers or common labor
Concrete	Cement Sand Gravel Water Transit mix Curing materials	Mixer Vibrator Power buggy Pump Crane Conveyor	Machine operator Common labor Cement finisher Foreman
Foundation wall Forming and stripping	Form panels or dimension boards Wire Ties Rods Nails		Foreman Carpenter Laborer
Reinforcing	Reinforcing steel Wire mesh Ties, chairs, and spacers		Foreman Iron worker Laborer

Table 11-B (cont.)

Operation	Supplies and Materials	Equipment (regular hand and power tools not listed)	Type of labor
Concrete			Foreman
	Cement	Mixer	Mixer engineer
		Vibrator	Laborer
	Sand	Power buggy	Cement finisher
	Gravel		
	Water	Pump	
	Transit mix	Conveyor	
		Crane	
Masonry units			Foreman
	Concrete block	Fork	Mason
		Lift	
	Light weight aggregate blocks	Crane	Tender
	Brick	Masonry saw	
	Stone		
	Ties		Laborer
	(Mortar)		
	Sand		
	Cement		
	Lime	Mixer	
	Masonry cement		
Openings and embedded items			Foreman
	Window frames		Carpenter
			Laborer
			Glazier
	Area way walls		Normally done with forming or reinforcing crew
	Lintels		
	Foundation vents		
	Door frames		
	Conduit		
	Pipe		
	Drains		
	Ducts		
Foundation drainage			Foreman
	Drain tile		Pipe layer
	Porous fill		Laborer
Waterproofing	Integral liquid repellent		
	Parging		
	Cement	Mixer	Foreman
	Sand		Plasterer
	Admixture		Tender
	Water		Machine operator
			Laborer

Table 11-B (cont.)

Operation	Supplies and Materials	Equipment (regular hand and power tools not listed)	Type of labor
	Asphalt		
	Coal tar pitch	Melter	
	Dry felt		Roofers
	Asphalt impregnated felt		
	Mastic		
Backfill	Water	Loader	Foreman
	Sand	Bulldozer	Machine operator
		Compactor	Laborer
		Compressor trucks	

ESTIMATING INFORMATION SHEETS

As an aid in estimating quantities of material, supplies, labor, hand and power tools, and equipment needed for various jobs, many estimating information sheets or guides have been developed over the years. The contractor will have access or will have developed a number of these.

QUANTITY ESTIMATING FOR CONCRETE

Accurate quantity estimating is important to success in the contracting business because it minimizes the expenses of manufacturing and hauling excess materials or the even higher cost of running short of concrete on the job. Estimating quantities of concrete is simply a matter of calculating the volume of space enclosed by the forms and adding a factor for losses and overages. Volume, of course, is computed by multiplying the length times the width times the height of the enclosed space. For simple shapes this can be done quickly and easily, but for buildings and complex shapes involving irregular or curved spaces (see Figs. 11-1 and 11-2 for formulas), the procedure becomes more involved.

I. Formulas Often Used in Estimating
Area of a square = length × breadth or height.
Area of a rectangle = length × breadth or height.
Area of a triangle = base × ½ altitude.
Area of parallelogram = base × altitude.
Area of trapezoid = altitude × ½ the sum of parallel sides.
Area of trapezium = divide into two triangles, total their areas.
Circumference of circle = diameter × 3.1416.
Area of a circle = square of diameter × 0.7854 = πr^2.
Area of ellipse = product of the two diameters × 0.7854.
Area of a parabola = base × ⅔ of altitude.

Contents of pyramid or cone = area of base × 1/3 altitude.
Volume of cylinder = $\pi r^2 h$.

II. Formulas Sometimes Used in Estimating

Area of a sector of circle = length of arc × 1/2 radius.
Area of a regular polygon = sum of its sides × perpendicular from its sides × perpendicular from its center to one of its sides divided by 2.
Surface of cylinder or prism = area of both ends plus length and × circumference.
Surface of sphere = diameter × circumference.
Area of surfacing sphere $(S) = \pi d^2$.
Volume of v $(V) = 1/6 \pi d^3$.
Diameter of sphere = cube root of solidity × 1.2407.
Diameter of sphere = square root of surface × 0.56419.
Circumference of sphere = square root of surface × 1.772454.
Circumference of sphere = cube root of solidity × 3.8978.
Contents of segment of sphere = (height squared plus three times the square of radius of base) × (height × 0.5236).
Contents of a sphere = diameter × 0.5236.
Side of inscribed cube of sphere = square root of diameter.
Side of inscribed cube of sphere = radius × 1.1547.
Surface of pyramid or cone = circumference of base × 1/2 of the slant height plus area of base.
Contents of frustum of pyramid or cone = sum of circumference at both ends × 1/2 slant height plus area of both ends.
Contents of frustum of pyramid or cone = multiply areas of two ends together and extract square root. Add to this root the two areas and × 1/3 altitude.
Contents of wedge = area of base × 1/2 altitude.

Figure 11–1

A number of shortcuts and tables have been developed over the years to expedite the quantity estimating job. In architectural drawings, dimensions are usually shown in feet and inches and even in fractions of an inch. Although it is possible to add and multiply such units, it is much easier to convert the inches and fractions to their decimal equivalents. Table 11-C shows inches and the corresponding decimal fractions of a foot. Dimensions such as these are much easier to add and multiply.

Table 11-C

Inches	Fractional Part of Foot	Decimal Part of Foot
1	1/12	0.08
2	1/6	0.17
3	1/4	0.25
4	1/3	0.33
5	5/12	0.42
6	1/2	0.50
7	7/12	0.58
8	2/3	0.67
9	3/4	0.75
10	5/6	0.83
11	11/12	0.92
12	1	1.00

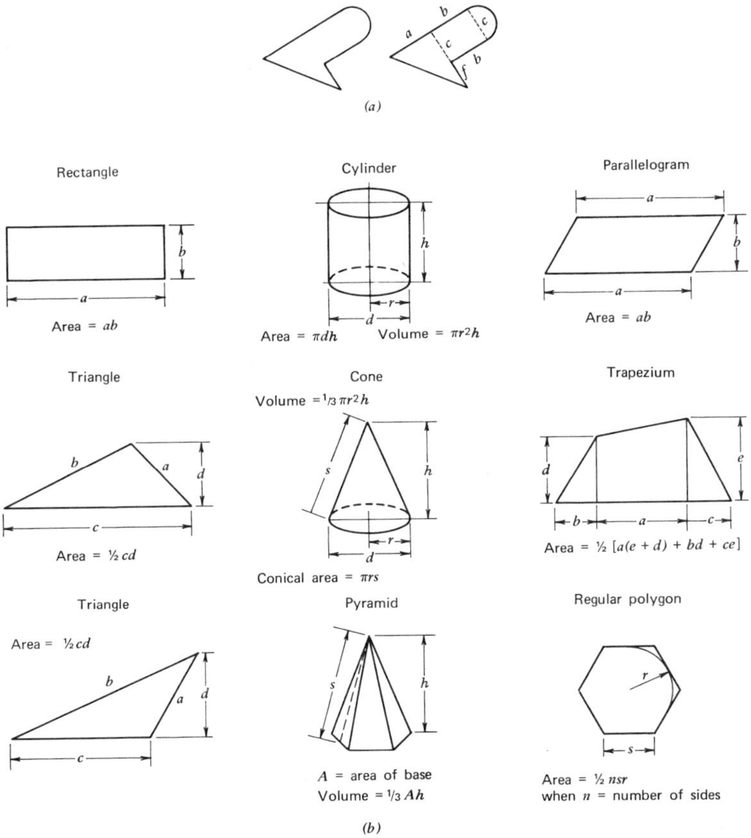

Figure 11.2 Concrete requirements for slabs of uniform thickness are calculated by simply multiplying the area of the slab by the thickness. Geometric formulas are used to determine the area of simple shapes. When an area has an irregular shape, however, it is usually easiest to divide the area into a group of regular shapes, as shown (*A*).

In estimating quantities for a concrete building or other complex structure, a table or form for listing each item is helpful. Columns should be provided for designation of the kind of member, its dimensions, unit of measurement, and its volume. Every single member must be listed to avoid underestimating quantities.

Dimensions should always be taken off the plans in the same order, namely, length, width, and height or thickness. This method avoids mistakes and makes it easier to check the work when necessary. Also, a definite order of following around the plan should be used. Start at the top left-hand corner and proceed in a clockwise direction; that is, work from left to right. Then, if one is interrupted, it is much easier to pick up the proper place again.

The method of listing quantities should conform with unit pricing and the system of cost keeping or accounting, since one is based on the other. For instance, if it is desired to price and to keep the cost of footings separate from that of foundation walls, the quantities must be listed separately.

To estimate the volume required in a given structural member, first multiply its length by its width or thickness by its height. This gives the volume of the member. If it is a footing going around the perimeter of a four-sided building, add the length of each side. Using this method, the volume of the corners is figured twice. This gives slight excess, but it is close enough for estimating purposes. This procedure is followed for each part of the building. When estimating members with windows or doors, be sure to deduct the volume they will occupy. When the total volume in cubic feet has been calculated, this quantity is divided by 27 to determine the number of cubic yards required.

Tables for determining the volume of various shapes of concrete members are available to minimize computations. For instance, tables give areas and perimeters of various shapes of columns. Others give areas of rectangular columns in square feet. Another gives the volume of conical column caps with various top and bottom diameters. Volume of concrete for stairs can be approximated by figuring the volume of the sloping slab when the dimension from the bottom of the slab to a perpendicular line bisecting the risers is used as the slab thickness (Fig. 11-3).

When all of the members of the building have been listed, the multiplications or extensions should be made. If these are all done together, time is saved and errors are avoided. To avoid mistakes in decimal points when extending, the figures multiplied together should be scanned and the product calculated mentally by using round numbers. For example, if $49 \times 72 \times 8$ is to be multiplied, the product might be mistakenly put down as 2822.4 instead of 28,224. Change the figures mentally to $50 \times 70 \times 8$, giving a product of 28,000, and the mistake is avoided. Several sheets can be gone over very quickly in this manner, relieving one of any uneasiness as to the correctness of the extensions. A check should also be made on additions by always adding a second time in the reverse direction. It is desirable to have an independent check made on all major quantities.

After all the totals are obtained, it is desirable to have a quick approx-

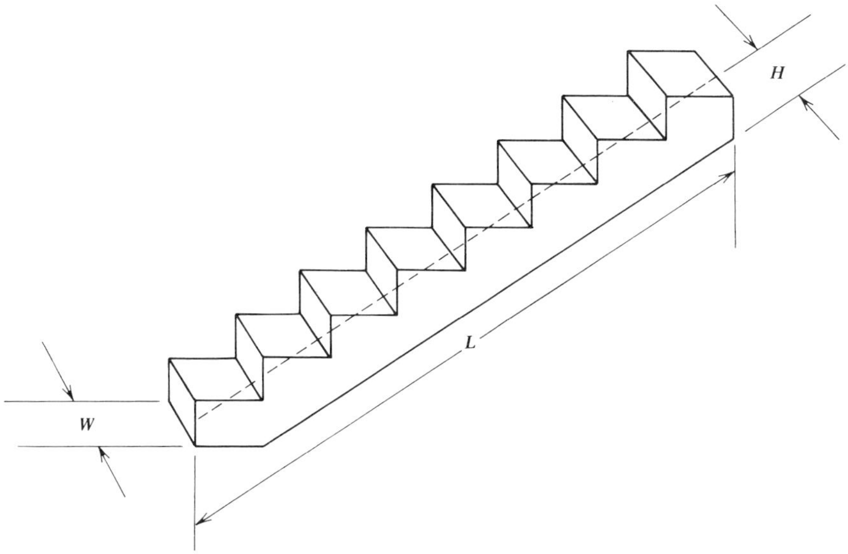

Figure 11–3 Figuring stair volume.

imate check on the figures. The total floor area of the building has already been calculated. Dividing the total concrete in cubic feet by this area gives the amount of concrete per supported area. An experienced estimator can tell at once by these units whether the totals are about right or if he has made a serious mistake, since for a particular type of building these units vary only within certain limits. Tables are also available for calculating the amounts of reinforcing steel used in concrete and for estimating the amounts of labor for mixing, transporting, placing, and finishing in many types of concrete construction.

STREET AND HIGHWAY ESTIMATING—CONCRETE

The approach to estimating paving for a street or highway project is different than for buildings. This is true for two reasons: first, in highway construction, especially when the job is large, a great deal of equipment and manpower is needed; and second, since the work is similar from day to day, the daily cost of the manpower and equipment is fairly constant.

Once you know your daily labor and equipment costs and how much pavement you can construct in a day, the total cost can be figured by determining the number of days the job will take, then multiplying this number of days by the daily labor and equipment, and finally adding the cost of materials.

Let's look at an example. Assume we are bidding on a residential paving job. The street is to be 36 ft wide including integral curbs on either side. The length of the project is 5000 ft. The total area to be paved with concrete is 20,000 sq yd and the pavement thickness is 5 in. We also have the following information:

Curbing: 6 in. × 6 in. requires 1 cu yd concrete/100 lineal ft.
Curing compound: $0.90/gal covers 200 sq ft/gal.
Transverse joints: sawed and sealed every 20 lineal ft.
Concrete cost: $19.50/cu yd.
Concrete production capacity: 278 cu yd/day for pavement.
18-ft-wide pavement can be placed in one pass and will require a sawed longitudinal joint.

The first step is to figure our daily production. We know we can't place more than 278 cu yd a day because that's all the concrete we can get. So let's see how much street we can put down with that amount.

We need to know how many square yards of pavement can be constructed with a cubic yard of concrete. First, we calculate the number of pavement thicknesses in 1 cu yd:

$$N = \frac{\text{thickness of a cubic yard in inches}}{\text{pavement thickness}}$$

$$N = \frac{36 \text{ in.}}{5 \text{ in.}} = 7.2 \text{ sy/cy @ 5 in. thick}$$

If we then multiply 278 cu yd by 7.2, we find that we can cover about 2000 sq yd with our daily supply of concrete.

To convert 2000 sq yd to lineal feet, we multiply by 9 (the number of square feet in 1 square yard) to give us the total number of square feet and divide by 18 (the width of pavement lane being placed):

$$\text{lineal feet per day} = \frac{2000 \times 9}{18} = 1000 \text{ ft}$$

This figure does not include the concrete for the curb, which requires an additional 1 cu yd for each 100 lin ft of pavement lane placed.

Next, we calculate the cost of the materials per square yard of paved surface.

Street Surface

$$\text{cost/sq yd of street surface} = \frac{\text{cost of concrete/cu yd} \times \text{no. of cu yd used}}{\text{no. of sq yd of street surface produced}}$$

$$\text{cost/sq yd of street surface} = \frac{\$19.50 \times 278}{2000} = \$2.711$$

Curbing on Each Side of the Street

$$\text{amount of concrete used} = \frac{1000 \text{ lin ft/day}}{100 \text{ lin ft/cu yd}} = 10 \text{ cu yd/day}$$

$$\text{cost of concrete} = \$19.50 \times 10 = \$195.00$$

$$\text{cost/sq yd of street surface} = \frac{\$195.00}{2000 \text{ sq yd}} = \$0.098$$

Total Concrete Cost/Sq Yd

cost of street surface	\$2.711
+ cost of curbing	0.098
	\$2.809
+ 4.0% overrun (0.04 × 2.809)	0.112
	\$2.921

Curing Compound

$$\text{cost of curing/sq yd} = \frac{\text{cost of compound/gal}}{\text{sq ft coverage/gal}} \times \text{sq ft in a sq yd}$$

$$\text{cost of curing/sq yd} = \frac{\$0.90}{200 \text{ sq ft}} \times 9 \text{ sq ft} = \$0.041$$

Sawing and Sealing Joints

transverse joint spacing = 20 ft

$$\text{number of transverse joints} = \frac{5000 \text{ ft}}{20 \text{ ft}} = 250$$

total length of transverse joints = 250 × 36 ft = 9000 ft

A longitudinal joint will be required in each 18 ft-wide paving lane.

total length of longitudinal joints = 2 × 5000 = 10,000 ft
total length of sawed joints = 19,000 lin ft
blade life = 3000 ft

$$\text{number of blades used} = \frac{19,000 \text{ ft}}{3000 \text{ ft}} = 6.33 \text{ or } 7 \text{ blades}$$

blade cost = \$300.00 each
\$300.00 × 7 = \$2100.00

sealer cost = \$0.20/lb and 25 lb of sealer is required for 100 lin ft of sawed joint.

$$\frac{19{,}000 \text{ ft}}{100 \text{ ft}} \times 25 \text{ lb} = 4750 \text{ lb}$$

$$4750 \times \$0.20 = \$950.00$$

blades and sealing material total = \$2100.00 + 950.00 = \$3050.00

$$\text{total/sq yd} = \frac{\$3050.00}{20{,}000} = \$0.153$$

The labor and equipment costs for one day, or 2000 sq yd, are itemized as follows.

Number of Men	Description	Pay Hour	Rate/ Hour	Labor/ Day	Equipment and Supplies: Rate/ Day	Equipment and Supplies
	Supervisor					
1	Foreman	8	\$10.00	\$ 80.00	\$ 17.00	½-ton pickup
	Forms					
2	Linemen	16	7.50	120.00	7.00	Miscellaneous tools, string, etc.
2	Grade men	16	7.50	120.00		
2	Place	16	7.50	120.00	30.00	6-in. road forms
1	Pin driver	8	7.50	60.00		
1	Truck driver	8	7.50	60.00		
1	Labor	8	7.50	60.00		
	Fine Grade					
1	Motor grader	8	10.00	80.00	96.00	18-ton diesel
½	End Loader	4	8.90	35.60	48.00	
½	Dump truck	4	7.50	30.00	45.00	
2	Template	16	7.50	120.00	7.50	
½	Roller	4	8.10	32.40	30.00	
½	Water truck	4	7.50	30.00	38.00	
	Place					
1	Dump man	8	7.50	60.00		
3	Puddlers	28.5	7.50	213.75	10.00	
2	Screed men	16	7.50	120.00	10.00	Vibrating screed and tools
2	Finisher	19	9.00	171.00	5.00	Straightedge, etc.
2	Finisher—labor	16	7.50	120.00	5.00	Burl and bridge, etc.
1	Finisher—curb	9.5	9.00	85.50		
1	Finisher—labor	8	7.50	60.00		
	Cure					
1	Labor	11	7.50	82.50	42.00	Comp. and spray equipment
	Joints					
1	Power saw	8	7.50	60.00	38.00	30 h.p. gasoline
1	Sealing equipment	8	7.50	60.00	17.00	Sealing equipment
	TOTAL			\$1,980.75	\$445.50	

The summary of all the estimated costs looks like this (note 5% has been included for office and job overhead and 15% has been added to the bid price for profit margin):

SUMMARY OF COST ESTIMATE

Total labor		= $1980.75
Insurance (workmen's comp., liability and soc. sec. 10% of 1980.75 labor)		= $ 198.08
Total equipment and supplies		= $ 445.50
Total labor and equipment		= $2624.33
Total labor and equipment and supplies / Daily production	= $2624.33 / 2000 sq yd	= $ 1.312
Concrete cost/sq yd		= $ 2.921
Curing material/sq yd		= $ 0.041
Sawing and sealing material for joints		= $ 0.153
Contingencies (moves, intersections, expansion joints, structure adjustments, etc.)		= $ 0.150
Total pavement cost/sq yd		= $ 4.577
Office and job overhead 5%		= $ 0.229
Profit 15%		= $ 0.687
Bid price/sq yd		= $ 5.493

$5.493 is our bid price per square yard. The total paving cost would then be 20,000 sq yd × $5.493 or $109,860.00.

Since 1000 lin ft, 18 ft wide would be placed each day, the paving work would require 10 working days since the street is 5000 ft long and 36 ft wide. This is figured according to the number of working days since inclement weather or other factors can interrupt paving.

CHAPTER 12
SCHEDULING

Scheduling in a *manufacturing* operation includes such jobs as planning, routing, scheduling, dispatching, and followup and is usually called production control. In the construction industry the same operations are performed, but because scheduling is the dominant activity, this is the title given to the procedure. Scheduling is one of the contractor's most important jobs. Often this phase is not done at all or is done in a haphazard fashion and, as a result, the contractor operates from crisis to crisis.

There are a number of methods and procedures that make scheduling easier for the contractor although no given method will fit every application. A system that works successfully for one contractor cannot be expected to operate well for another without major changes.

The scheduling function produces nothing itself—it merely programs or plans for things to be done in a certain pattern at predetermined times. Since the scheduling function produces no direct profit, the procedures must be carefully controlled. Once they have realized some of the benefits of good scheduling, contractors too often go overboard for scheduling techniques and charting devices and spend more money and time than is justified. Expenses of maintaining the function can be recovered in only two ways: from money saved by good scheduling or by adding the cost of scheduling to the owner's or customer's purchase price.

Since the second method is obviously undesirable, the system should be just as simple as possible so that it yields no unused information or introduces no unduly complicated control measures.

The first step in scheduling is to determine the completion date of the structure, or in the case of the small homebuilder, a schedule for the completion of each house that he is constructing. When the completion dates have been determined, the next step is to determine the construction methods and the number and size of the work crews that will be necessary to meet these completion dates. The final construction schedule depends on knowledge of the best sequence for each phase of construction, and the time needed for the crew to complete each job.

In practice a crew of men is usually assigned to each job. Finding the number of hours it takes a crew to do a job is not a simple matter of dividing the total man-hours by the number of men on the crew. Several factors must be considered. The question of the most efficient size of crew immediately arises. Many builders claim that a moderately large crew size makes for greater worker productivity, because of the competition between workers. If the crew is too large, productivity may suffer because of the inability to provide enough supervision. Some builders feel that small crews are better because productivity and quality are higher. Once the optimum crew size has been decided, the number of man-hours of work that the crew can complete in a day should be estimated. When this crew is assigned to a job, the amount of time it will take to complete that job can be figured.

Construction scheduling, then, is using the time estimates established in the planning stage to determine beginning and ending dates for each construction operation. This procedure is not especially complex for the small builder or for the simple construction job. In simple scheduling cases the builder would complete each operation singly and consequently with no overlapping of any operations. The resulting schedule would resemble the one shown in Figure 12-1.

This sort of scheduling is not well adapted to the needs of most builders. Most builders would want to complete at least certain operations concurrently, that is, overlap them in terms of time. This sort of operation is illustrated in Figure 12-2.

Concurrent scheduling has several advantages: the overall time to complete construction is reduced, greater flexibility in scheduling is possible, and one operation can be moved ahead or back without delaying the entire schedule. This type of scheduling does complicate the charting

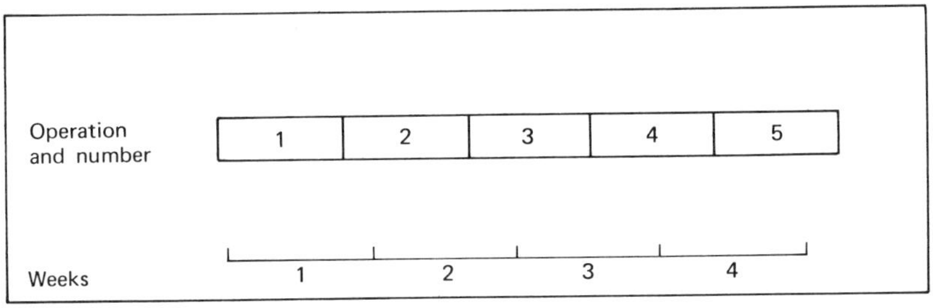

Figure 12–1 Charting single and consecutive operations.

process, however, especially when a number of jobs are being completed simultaneously. There are several ways to handle concurrent scheduling efficiently. One method is to use a construction schedule form, as in Figure 12-3. In this form the builder first numbers the sequence of operations in the exact order he expects to complete them. If two or more are to be completed concurrently, the exact sequence of numbering may not be too significant. The builder then enters the standard time for each operation in man-hours as determined in the planning stage. This figure is then converted to crew-days, that is, the number of days it will take the crew to complete the assigned job. Using this information, he enters the starting and completion dates for each operation. This schedule will give him a graphic view of the entire scheduling of the job. He must carefully consider which portions of the job must be completed before subsequent operations can begin.

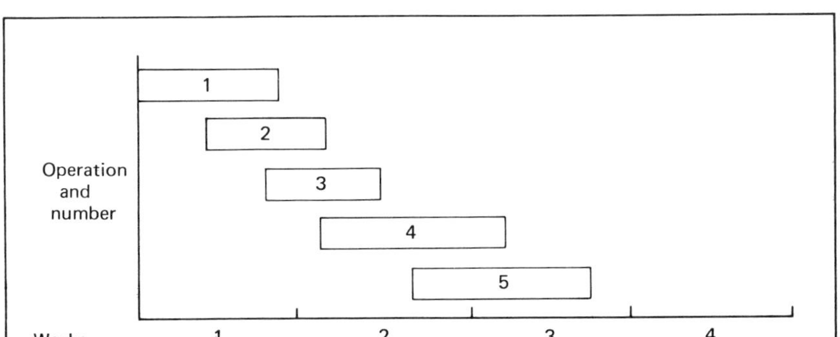

Figure 12–2 Charting of operations scheduled concurrently.

Figure 12–3

Operation	Standard Time	Allotted Crew Days	Subcontractor	Percentage Completed					Behind Schedule (check)	Record Beginning Days of Weeks
				20	40	60	80	100		
Land acquisition										
Survey										
Design										
Permits										
Layout										
Excavation										
Footings										
Foundation										
Waterproofing										
Back fill										
Rough floor										
Exterior walls etc.										

CRITICAL PATH METHOD

To help him keep track of overlapping jobs, another method of scheduling has been devised called the critical path method or CPM. This planning technique is being used by construction companies to cut their job time, sometimes 20% or more, and shows them how to boost profits by getting the job done in the shortest possible time. It is a graphic method of scheduling that utilizes arrows and symbols to plan the progress of a job.

CPM is no more mysterious than the bar chart (Fig. 12-2) used almost universally by contractors to schedule their work. The bar chart has a bar for every job in the operation. CPM uses an arrow on a diagram for every job. But where a bar chart calls for a lot of reading between the bars to figure where each job stands in relation to the others and where it should stand at any given date, a CPM diagram shows visually the exact interrelationship of every job that has to be done to complete the operation—whether there are 10 or 10,000 jobs.

The name, critical path method, grows out of a unique advantage of the system. It shows graphically exactly which jobs—and they seldom total more than 10% of the total jobs in an operation—are critical; that is, which jobs, if delayed, will delay the whole operation. These are the jobs that must be tightly scheduled and controlled, that must be completed on time, and that must be speeded up if the schedule must be speeded up. You'll see why on the following pages.

With CPM, it is harder to overlook a small but critical operation in planning. Making the inevitable scheduling changes after the job is underway is easier. The CPM diagram gives subcontractors a better picture of how their jobs relate to others. It gives them an incentive to schedule their work better and can result in low subcontractor prices. CPM saves money and avoids delays in delivery. It assures that materials are ordered and delivered on schedule, but not too far in advance. And, perhaps most important of all, CPM diagrams give management information on the whole job at a glance.

Some contractors use computers to build a critical path diagram and set up the best schedule. So CPM has developed an aura of higher mathematics and complexity. But on the scale that builders might use it—for houses, apartments, and light commercial work—CPM is a pencil-and-paper process.

Drawing a critical path diagram—like solving an algebra problem or tying your shoes—is easier to do than to explain. But the only way to learn is to begin.

The first step is to list all the jobs that have to be done to complete the operation. Let's say there are 10—*A*, *B*, *C*, *D*, *E*, *F*, *G*, *H*, *I*, and *J*.

In listing the jobs for a house, the breakdown must be most detailed. For example, the list might begin: clear site, layout site, prepare footings, order footing inspection, order footing concrete, pour footing, order block, build foundation wall, and so on. Major jobs—like framing a two-story house—might well be broken down into two jobs: first-floor framing and second-floor framing (yet all framing might be listed as one job if the work does not have to be interrupted for another trade to do a job). All deliveries of materials and projects must be listed as jobs. The list must be made by someone completely familiar with the building process.

When the job list is ready, the arrow diagram can be drawn. The first step: determine which comes first. An arrow is drawn representing this job. The tail of the arrow represents the start of the job; the head, the end of the job. The length of the arrow is labeled with the name of the job (*A*) (Fig. 12-4).

The next step is to see what jobs, if any, can be started concurrently with *A*. Let's say that none can.

The next step is to see what jobs can be started when *A* is completed. Say there are two—*B* and *C*. An arrow representing each of these jobs is drawn—starting with the head of arrow *A* (Fig. 12-5).

The next step is to draw arrows for any jobs that can be started as soon as job *B* is complete. Say there are three—*D*, *F*, and *H* (Fig. 12-6).

What jobs can be started as soon as job *C* is complete? Only one—job *I* (Fig. 12-7).

The same process, then, is continued for every task in the operation. And the questions are the same for each arrow added to the diagram: (1) What jobs must be finished before this job can begin? (2) What jobs can be accomplished concurrently with this job? (3) What jobs cannot begin until this job is finished?

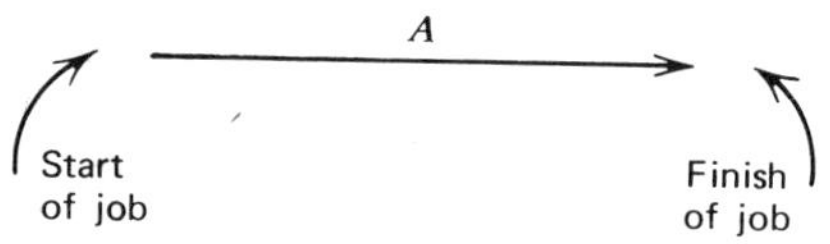

Figure 12–4

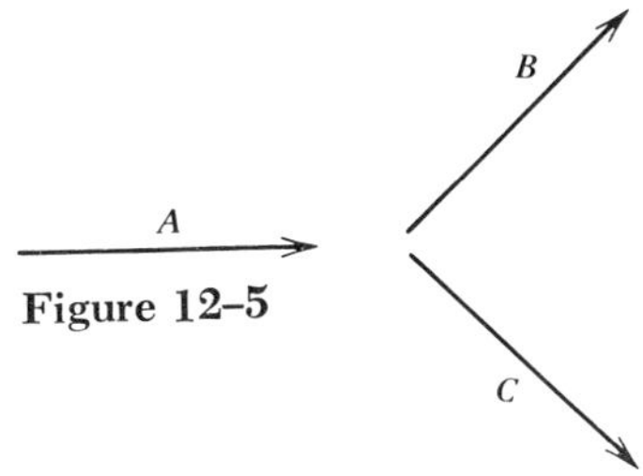

Figure 12–5

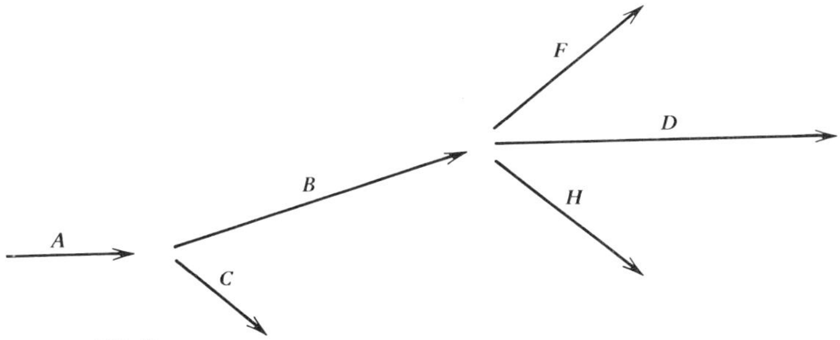

Figure 12–6

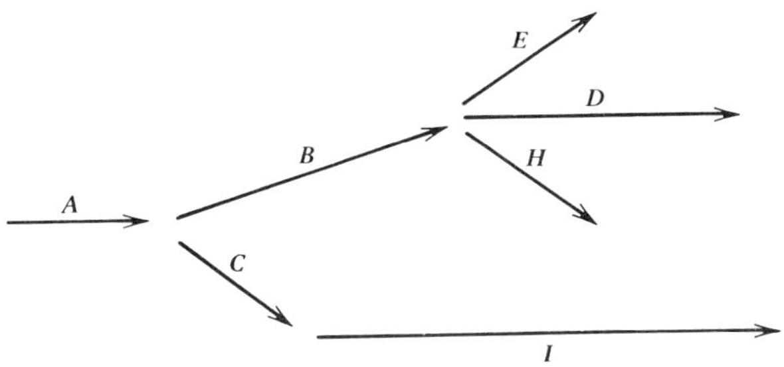

Figure 12–7

At some point in the drawing of the diagram, there will be jobs that cannot be accomplished until two or more other jobs are completed. For example, say that job *J* cannot be started until both *H* and *I* are finished. The arrows for *H* and *I* are extended to a single point, and the arrow for job *J* is drawn from there (Fig. 12-8).

Finally, let's say that the diagram is complete except for job *G*. Job *G* cannot be started until all other jobs are complete. Again, the heads of

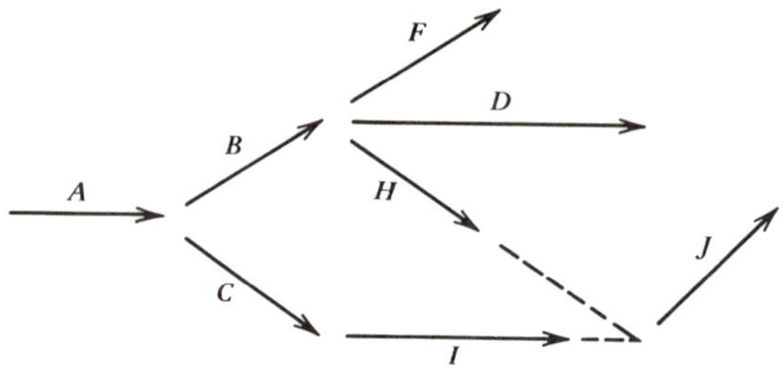

Figure 12–8

all arrows are extended to a single point, and the arrow for job *G* drawn (Fig. 12-9).

The final step—redraw the diagram for clarity. (In drawing an actual diagram, it may be necessary to redraw several times along the way to avoid crisscrossing lines.) Note that although arrows (here, arrow *F*) can be bent, this has no significance (Fig. 12-10).

The arrow diagram used for an 18-unit apartment house has well over 50 arrows (Fig. 12-17). An arrow diagram for a major construction job often has thousands of arrows and must be printed on a sheet of paper stretching around two sides of a room.

Usually, simply to eliminate the need for printing the name of each job along the arrow representing it (which is easy enough to do if the name of the job is *A*, but much harder if the name of the job is "install perimeter footing"), the arrow diagram is numbered. The tail of the first arrow —the start of the job—is labeled 0, and the head of the first arrow is labeled 1, and so on. These numbers are usually circled to avoid confusion with other numbers placed on the diagram later. In the example below, job *A* becomes 0-1; job *B*, 1-2; job *F*, 2-6; and so on (Fig. 12-11).

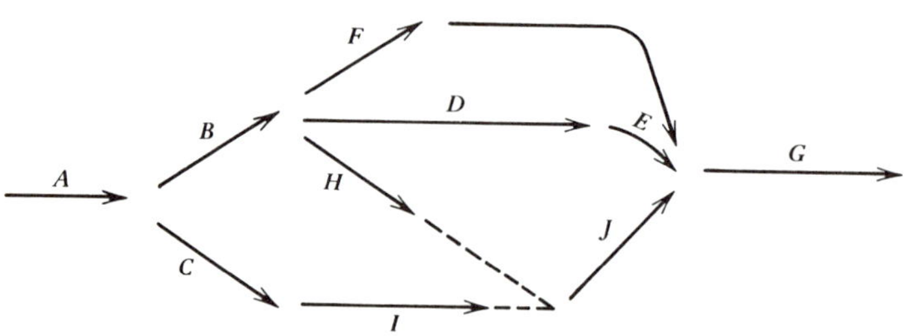

Figure 12–9

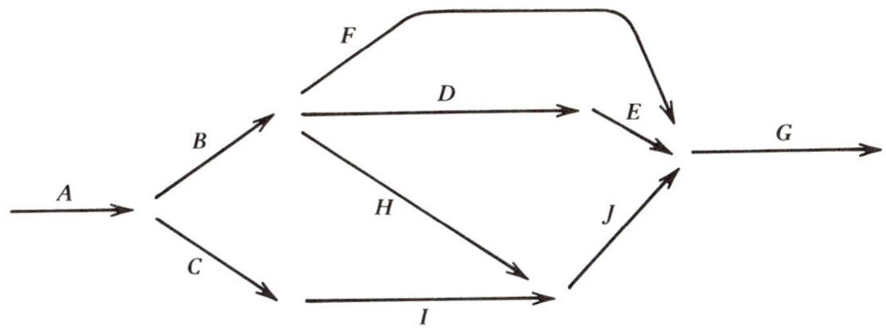

Figure 12–10

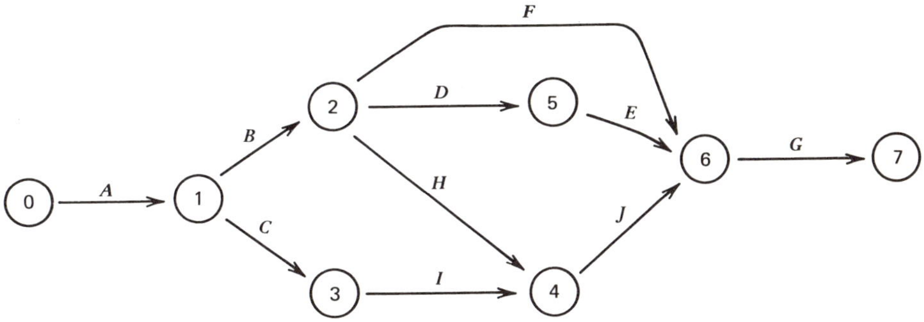

Figure 12–11

The next step is to allot a time (usually in days) to each job arrow. This time (and here again is where an experienced man comes in) is the best possible estimate of the amount of time that the job will take, based on crew size, delivery schedules, possible bad weather, and so on. This time is noted at the midpoint of each arrow on the diagram. Below, for example, job *A* (job 0-1) has been estimated at 3 days; job *B* (job 1-2) has been estimated at 2 days; and so on (Fig. 12-12).

At this point the critical path can be drawn. The process (not unlike those puzzles on the comic page where you find your way out of the maze) is to trace the longest path in time through the diagram. This is a trial-and-error process. For example, tracing along the dashed line, below, and adding up the time estimates (the numbers at the middle of each arrow), it is 13 days from start to finish (Fig. 12-13).

But tracing along the heavy line, below, the time estimates add up to 20 days—the longest possible path in time. This is therefore the critical path (Fig. 12-14).

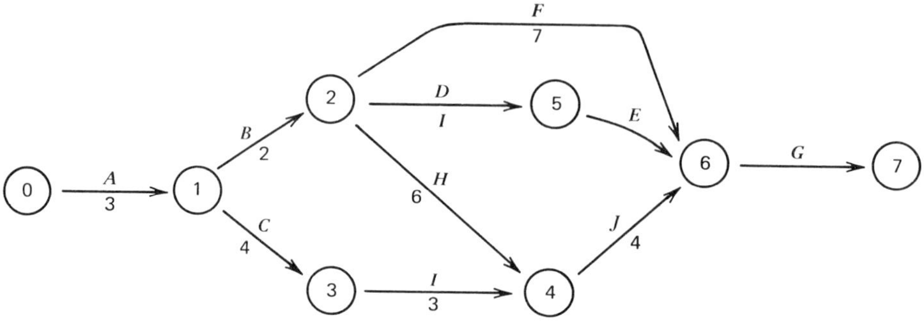

Figure 12–12

The jobs along the critical path are the critical jobs—the jobs that determine how long the operation will take from start to finish.

This is the area where a critical path diagram is much more useful than a bar chart. Here is the bar chart for the same 10-job project, to be completed in the same 20 days. It is not possible to find from this chart which jobs are critical and which are not (Fig. 12-15).

The next step is to trace along the critical path and note at the head of each arrow the day on which each of these critical jobs can be completed. These days are usually put in a square. On the drawing below, job *A* can be completed on day 3, job *B* on day 5 (3 days for job *A*, 2 days for job *B*), and so on. The final figure—20—at the head of the last arrow on the critical path is the number of days it will take to complete the operation according to estimates (Fig. 12-16).

There is only one more thing to be determined: the so-called float time for jobs that are not on the critical path. The simplest example in the

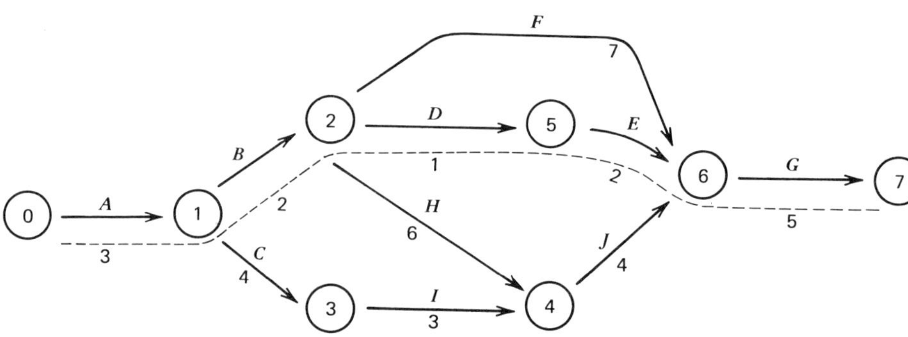

Figure 12–13

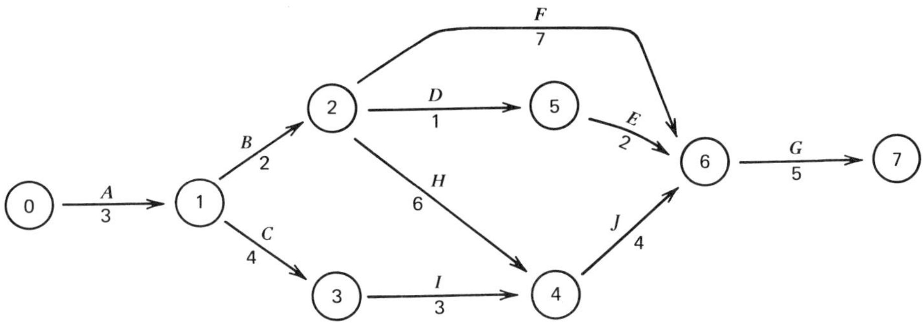

Figure 12–14

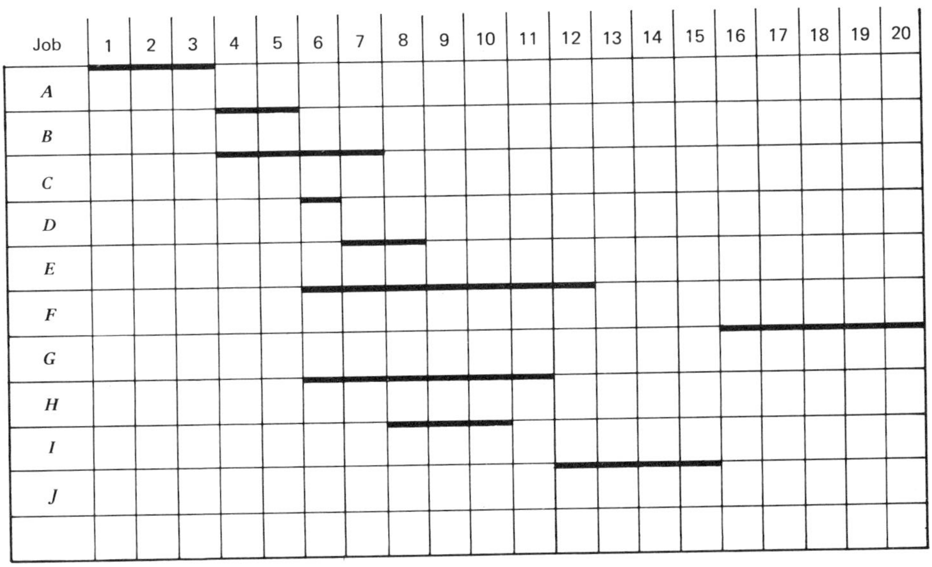

Figure 12–15

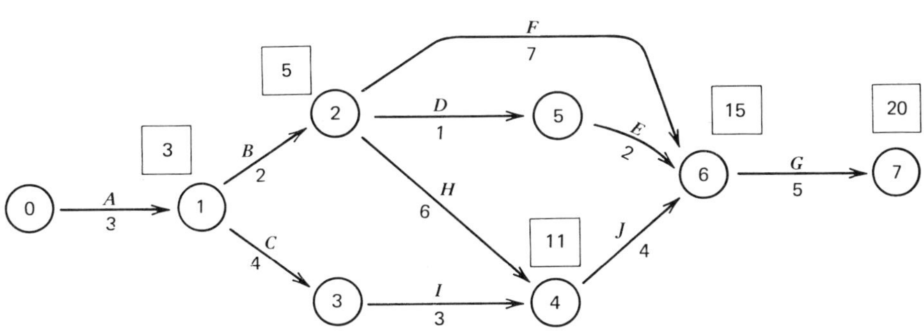

Figure 12–16

diagram is job *F*. Job *F* takes 7 days. It can be started as soon as job *B* is finished—on day 5. But it need not be finished until day 15 (see the head of arrow *F*). There is a 10-day period in which this 7-day job can be done. So its float time is 3 days. Similarly, jobs *D* and *E* together have a float time of 7 days; they can be done anytime during a 10-day period, and together they take 3 days to do. This float time can be assigned all to one or the other, or divided, or assigned to both as a pair of related jobs.

Float time can be used to adjust the schedule to fit crews and equipment available and to deploy men and equipment around a project most efficiently. It may shrink if the overall time schedule is shortened.

Critical jobs, by definition, have no float time.

The usual practice in building is to establish a predetermined finish date for a project. The critical path method is, of course, a sound way to establish a realistic finish date. And it is a most valuable management tool when—for whatever reason (bad weather, delivery delays, etc.)—the total project time must be shortened.

With normal scheduling methods, the traditional approach is to put all jobs on a crash program. Extra money, men, equipment, and materials are poured into every job in the project.

But under CPM scheduling, since you know which jobs are critical, only these jobs need be speeded up—and this can mean a big cost saving. This speedup can be started at any time during the course of the project. When the critical path is shortened, some of the noncritical jobs may not have enough float time, so they too must be speeded up (Fig. 12-17).

EXAMPLE OF CPM SCHEDULING

Now let's draw a critical path diagram for the foundation we estimated previously.

The first step is listing the job number, description, float, and days estimated for completion. The description and completion time can be filled out before the diagram is drawn, and the job number and float can be entered after the diagram is finished.

Draw an arrow diagram of the jobs, paying attention to the starting and completion times, the order in which work can be done, and concurrent jobs.

You will probably have to redraw the diagram several times before you are satisfied with it.

Enter your estimated times as you work along.

Plot the critical path and total the time for the entire job.

Job Number	Description	Float Days	Estimated Time Days
0-1	Layout		0.50
1-2	Excavation		1.00
2-3	Footings, main, forming		1.00
2-4	Footings, post, forming	4.00	0.50
3-5	Footings, main, place, and reinforce		2.00
4-8	Footings, post, place	4.00	0.50
5-6	Foundation, main, erecting forms		4.00
6-7	Foundation, main, place, and reinforce		1.00
7-8	Foundation, main, strip forms		1.00
8-9	Foundation, waterproofing	0.50	1.00
8-9	Windows, areaways, frames, glazing		1.50
8-17	Floor slab, placing and finishing	4.00	1.00
9-10	Foundation, backfilling		0.50
10-11	Excavation, porches		0.50
11-12	Footings, porches, forming		0.50
12-13	Footings, porches, placing		0.25
13-14	Foundation, porches, forming		1.00
14-15	Foundation, porches, placing		0.25
15-16	Floors, porches, placing, and finishing		0.50
16-17	Foundation, porches, strip forms		0.50
17-18	Clean up		0.50

COMPUTER-ASSISTED SCHEDULING

On many projects, especially large ones, a computer can be used to advantage in critical path planning. It can investigate the many paths through a project and the extremely large number of choices for speeding up a project that are available. This permits the builder to develop a schedule that best suits the needs and requirements of any set of circumstances. A computer programmer can take the data that the contractor has developed, that is, his arrow diagram estimates of time and costs and his float time estimates, and convert them to machine language. Many computers are already programmed to handle the critical path method and many are set up to handle the PERT (Project Evaluation and Review Technique) system, a variation of CPM that is extensively used by the government. A programmer will know how to adapt a computer that is programmed to handle PERT so that it will process CPM data. When the machine has received the contractor's time and cost data, it will develop a schedule for the project and store it on magnetic tape. The computer then considers whether a schedule of shorter duration could be produced. If so, it prints a schedule summary that shows the relationship of direct project costs and project duration (Table 12-A).

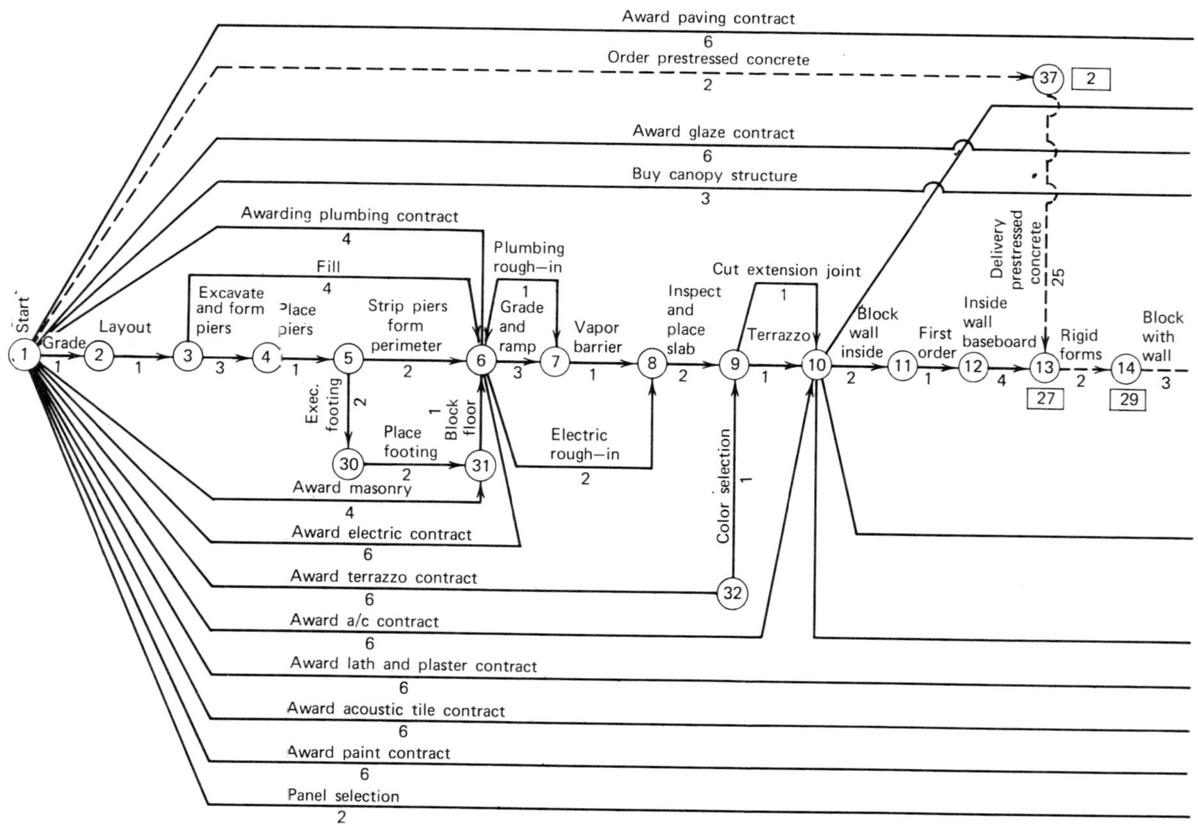

Figure 12–17 This is the CPM diagram for an 18-unit apartment.

Look complicated? It isn't. It's just longer (63 jobs) than the 10-job hypothetical example gone through previously. The dashed line is the critical path. The operation will be complete on the 74th working day (see boxed number at end). Notice the second job on the critical path (job 37-13, delivery of prestressed concrete)

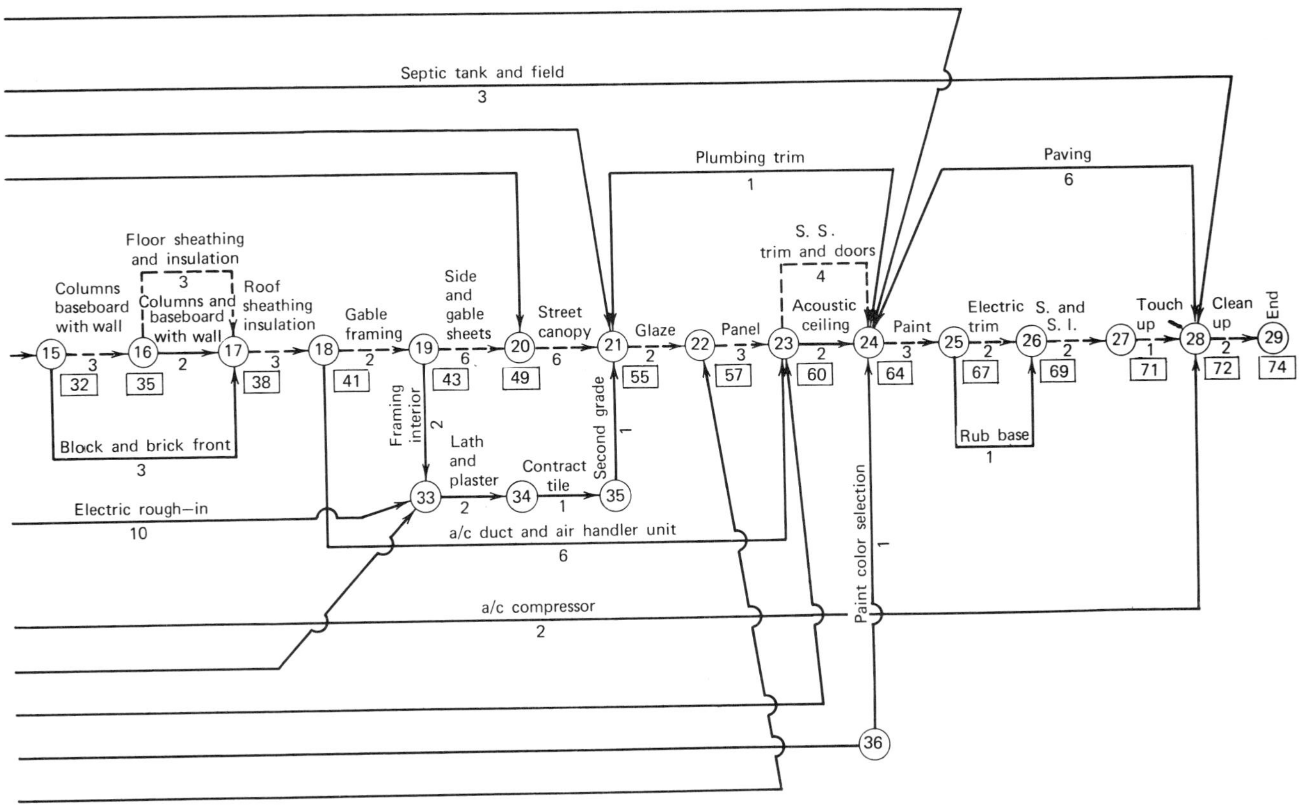

is given 25 days, must be finished by day 27. Could the job be shortened by asking for faster delivery? Not much, because the next longest path (1-2-3-4-5-30-6-7-8-9-10-11-12-13) through the diagram totals 26 days; these 14 jobs together have a total of only 1 day float time.

Table 12-A
GE 225 CPM SCHEDULE SUMMARY
SPECIAL SHORT SUMMARY

XYZ PROJECT 31 January 1969

Schedule Number	Project Duration	Direct Cost	INDIRECT COST	Total Cost
0	141	42338		
1	137	42388		
2	135	42575		
3	125	42850		
4	121	43155		
5	118	43960		
6	112	45000		

This summary is analyzed by the project manager to find out which of the several schedules should be used. The schedule summary has supplied information about the relationship between project duration and direct costs. Indirect costs can include overhead, loss of market, penalty clauses, and bonus clauses.

The indirect costs for a project are calculated and added to the direct costs so that a total cost is formed for each choice. The schedule with the least total cost is the one that should be chosen.

In the example in Table 12-B indirect costs included were only normal overhead at $100 a day. Schedule 4 is the one with the least total cost, $55,255. This indicates that a project manager could not only finish his project 20 days early (in 121 days rather than 141), but also save about $1200 in the process.

WHAT DOES THE COMPUTER PROVIDE THE CONSTRUCTION MANAGER?

When the project manager has chosen the schedule to be used, the computer center is informed of the choice and is instructed as to the date the project should start, the holidays to be observed, any weather days to skip, and the number of days in the work week. With this information the computer selects, from the magnetic tape containing all the schedules, the desired one and prints it as a calendar-dated schedule. The schedule provides information about the following:

1. Status of each activity in the project—whether it is critical or not.
2. Duration scheduled for each activity in days and in weeks and days.
3. Scheduled cost of each activity.

Table 12-B
GE 225 SCHEDULE SUMMARY
SPECIAL SHORT SUMMARY

XYZ PROJECT 31 January 1969

Schedule Number	Project Duration	Direct Cost	INDIRECT COST	Total Cost
0	141	42338	14100	56438
1	137	42388	13700	56038
2	135	42575	13500	56075
3	125	42850	12500	55350
4	121	43155	12100	55255
5	118	43960	11800	55760
6	112	45000	11200	56200

4. Earliest start and finish date for each activity.
5. Latest start and finish date for each activity.
6. Scheduled start and finish date that considers for each activity the amount of float that has been allocated.
7. Amount of float available and scheduled for each activity.
8. Date the project will finish if the schedule is met.

With the selected schedule, the project manager is in a position to begin his project. He will keep a close check on the progress of all activities. The computer center will have kept a magnetic tape containing all the details of his schedule. Periodically, during the life of the project, the computer can be supplied with information about the progress of activities and can produce status reports based on the actual job data. This is the procedure involved in the control-monitor phase. The information supplied about activities can include instructions to add a new activity, to delete an activity, or to change the duration on scheduled finish date of an activity, as well as the date on which an activity was started and ended.

Using this data, the computer produces a report that indicates the expected project finish date compared with that on the schedule; the status of the project compared to the schedule; activities overdue; the actual "used" duration for each activity compared with scheduled duration; the amount of float remaining for an activity; and whether an activity is early or late and by how much. With this information a manager can tell how the project stands at all times.

In addition to the status report, another report is produced showing which activities are the most crucial based on the amount of float available. Five separate paths of varying criticalness are listed. The first path provided is the critical path—the path with the least amount of float. The second is that set of activities closest to criticalness, the path with the next least float, and so on through five levels. This is done so that the project manager can keep a close watch not only on the critical activities, but also on activities that could become critical.

The amount of computer time required to obtain all necessary schedules and reports is rather small. For this reason it is feasible to use the services of a computer center from which information can be supplied by mail.

SECTION THREE

COMPILATION AND EVALUATION OF DATA

CHAPTER 13
ORGANIZATION AND PRESENTATION OF DATA

COLLECTION OF DATA

There are two general purposes behind the collecting of data, and in many cases both reasons play a part. One is concerned with the presentation of information, represented by the data, in such a way that it can be readily and accurately comprehended. This presentation usually involves the use of certain numerical characteristics of the data, called *statistics* (singular, *statistic*), to convey effectively and efficiently various features of the data. *Sampling* is the second reason for collecting data. We want to be able to generalize, according to established methods, from the collected data to a more comprehensive whole that includes the data. In most cases, testing an entirety of items on some property of interest, such as compressive strengths, for example, is impracticable if not impossible. Instead, we have recourse to what is called *sampling*. We take a *sample* of items and use the resulting data to make inferences regarding properties of a larger class from which the sample was selected. This whole class, about which conclusions are to be drawn, is called a *population.*

For instance, an investigator may be interested in the compressive strength of a population of concrete. This population must be clearly defined at the outset of any investigation; it may, for example, be a particular batch of concrete, all the concrete placed on a certain day for a job, or all the concrete for a given structure. Once the population is clearly established, the investigator would take a sample of concrete cylinders from this population and test their strengths. These strength measures would vary among themselves and collectively constitute the data.

If this investigator drew additional samples of concrete test cylinders from the population, we should not expect these samples to be identical; indeed, probably no two samples would be the same. This is because of variations and inaccuracies inherent in all natural phenomena, including manufacturing operations. Because of this variation among samples, sampling can never yield certain indications of population characteristics. Certainty about an underlying population is not possible on the basis of sampling. Instead, any conclusions are, at best, only *probably* correct. For this reason, all statistical inference is presented in terms of statements of probability.

As an illustration, suppose the investigator in our example is interested in the average strength of the population of concrete. He would then figure the average strength for the data of the sample. This average is an example of a statistic and would be used together with other statistics to infer a value for the population average. This conclusion, however, would be no more than an "educated" guess. Much more will be said about inferences based on sampling in Chapter 15.

Numerical characteristics of populations, such as the population average cited above, are called *parameters*. As we have seen, the corresponding characteristics for samples, computed from actual data, are called statistics. The actual or true value of a population parameter is generally not known.

Statisticians have found it convenient to use different symbols for the same characteristics of samples and populations. Population parameters are indicated by lower-case Greek letters (such as μ or σ) and statistics by ordinary upper- or lower-case) English letters ($\overline{X}$, SD).

A consideration of the utmost concern during sampling is that the sample be truly *representative* of the population from which it is taken. We would like to be sure that the sample is a miniature picture of the population that it is supposed to represent. The most common technique designed to achieve this is called *random sampling*. This means that the sample is chosen in such a way that every item in the population has an equal chance of being chosen throughout the sampling operation. The theory is that by drawing such a sample all aspects of the population should be represented in the sample.

Just how random sampling is done in practice depends on the particular case, and obtaining a random sample is not generally as easy as it might seem. Here is an example. If we were interested in the quality of ready mixed concrete being regularly delivered to a job site, we would not want to cast all test cylinders from the discharge of a single truck. Moreover, we would not want test cylinders to be taken only during the initial discharge of concrete from the trucks. Instead, we would want some way of randomly picking the trucks to be sampled and would also want to take concrete specimens at different points during discharge. It is important to appreciate and remember this requirement of representativeness of a sample.

An even more fundamental requirement of data that is to be useful statistically is that it be obtained under *controlled conditions*. Basically what this means is that the data be homogeneous. In order for this to be true, the measurements constituting the data must be made on material produced under essentially the same conditions, and test conditions must be essentially the same as well. For instance, if we were interested in the quality of concrete block being produced by a manufacturer, we would *not* want to indiscriminately include in our sample block cures by dissimilar methods, for example, atmospheric pressure steam curing and autoclaving. Furthermore, we would want to test each block in an essentially identical way according to standardized testing procedures.

If a set of data contains parts that were obtained under different test conditions, or that represent material produced under different conditions, the data should be divided into separate subgroups and these subgroups treated independently. Treating such data as a homogeneous whole frequently leads to results of little practical value.

FREQUENCY DISTRIBUTIONS

Let us assume, then, that we have obtained a set of measurements on some property of a sample of items and that these measures are homogeneous. Furthermore, we shall assume that the sample was obtained so that it can be considered representative of a population that includes the sample. Without an intelligent treatment of this data, worthwhile interpretation is practically impossible. Let us now take an actual example and consider ways of arranging and condensing the data into a form appropriate for practical use.

Table 13-A
ORIGINAL DATA—TRANSVERSE STRENGTHS OF
270 BRICKS OF THE SAME TYPE (measured to the nearest 10 psi)*

860	1320	820	1040	1000	1010	1190	1180	1080	1100	1130
920	1100	1250	1480	1150	740	1080	860	1000	810	1000
1200	830	1100	890	270	1070	830	1380	960	1360	730
850	920	940	1310	1330	1020	1390	830	820	980	1330
920	1070	1630	670	1150	1170	920	1120	1170	1160	1090
1090	700	910	1170	800	960	1020	1090	2010	890	930
830	880	870	1340	840	1180	740	880	790	1100	1260
1040	1080	1040	980	1240	800	860	1010	1130	970	1140
1510	1060	840	940	1110	1240	1290	870	1260	1050	900
740	1230	1020	1060	990	1020	820	1030	860	850	890
1150	860	1100	840	1060	1030	990	1100	1080	1070	970
1000	720	800	1170	970	690	1020	890	700	800	1150
1140	1080	990	570	790	1070	820	580	820	1060	980
1030	960	870	800	1040	820	1180	1350	1180	950	1110
700	860	660	1180	780	1230	950	900	760	1380	900
920	1100	1080	980	760	830	1220	1100	1090	1380	1270
860	990	890	940	910	1100	1020	1380	1010	1030	950
950	880	970	1000	990	830	850	630	710	900	890
1020	750	1070	920	870	1010	1230	780	1000	1150	1360
1300	970	800	650	1180	860	1150	1400	880	730	910
890	1030	1060	1610	1190	1400	850	1010	1010	1240	
1080	970	960	1180	1050	920	1110	780	780	1190	
910	1100	870	980	730	800	800	1140	940	980	
870	970	910	830	1030	1050	710	890	1010	1120	
810	1070	1100	460	860	1070	880	1240	940	860	

* Data from ASTM *Manual on Quality Control of Materials*, 1951, page 3.

Table 13-A contains data for the transverse strengths of 270 bricks of a single brand. These strength measures have been recorded to the nearest 10 psi in the order in which they were obtained. Since it is very difficult to make sense out of such a jumble of numbers, what is usually done as a first step in treating data is to arrange the numbers in an increasing order of magnitude. This arrangement is shown in Table 13-B. An arrangement of data, such as that in Table 13-B, that indicates the frequency of each of the possible measurements is termed a *frequency distribution.* In particular, an arrangement such as that of Table 13-B is called an *ungrouped frequency distribution*, as distinguished from the grouped frequency distribution discussed below.

The ungrouped frequency distribution of Table 13-B gives some information that is not obvious from the original data of Table 13-A. For instance, we can tell the minimum and maximum values and also form some idea of the relative frequencies of the different values, giving us a

Table 13-B
UNGROUPED FREQUENCY DISTRIBUTION OF THE DATA OF TABLE 13-A

270	780	830	870	920	970	1020	1070	1100	1180	1310
460	780	830	880	920	980	1020	1070	1100	1180	1320
570	780	830	880	920	980	1020	1070	1100	1180	1330
580	790	840	880	920	980	1020	1070	1100	1180	1330
630	790	840	880	920	980	1020	1070	1110	1180	1340
650	800	840	880	930	980	1020	1070	1110	1180	1350
660	800	850	880	940	980	1020	1070	1110	1180	1360
670	800	850	890	940	990	1030	1080	1120	1190	1360
690	800	850	890	940	990	1030	1080	1120	1190	1380
700	800	850	890	940	990	1030	1080	1130	1190	1380
700	800	860	890	940	990	1030	1080	1130	1200	1380
700	800	860	890	950	990	1030	1080	1140	1220	1380
710	810	860	890	950	1000	1030	1080	1140	1230	1390
710	810	860	890	950	1000	1040	1080	1140	1230	1400
720	820	860	890	950	1000	1040	1090	1150	1230	1400
730	820	860	900	960	1000	1040	1090	1150	1240	1480
730	820	860	900	960	1000	1040	1090	1150	1240	1510
730	820	860	900	960	1000	1050	1090	1150	1240	1610
740	820	860	900	960	1010	1050	1100	1150	1240	1630
740	820	860	910	970	1010	1050	1100	1150	1250	2010
740	820	870	910	970	1010	1060	1100	1160	1260	
750	830	870	910	970	1010	1060	1100	1170	1270	
760	830	870	910	970	1010	1060	1100	1170	1270	
760	830	870	910	970	1010	1060	1100	1170	1290	
780	830	870	920	970	1010	1060	1100	1170	1300	

rough picture of the distribution. However, even this form is difficult to take in at a glance since there are so many numbers. For this reason, when the number of data items is sufficiently large (say, 25 or more) a more compact form is preferable.

GROUPED FREQUENCY DISTRIBUTIONS

A more compact form of a frequency distribution can be obtained by bringing together the single measures into a number of groups, each group containing an equal number of measurement units. This has been done for the data on brick strengths in Table 13-C, where the size of each group has been made 150 psi. The column on the right gives the frequency (f) of occurrence for each of the groups, that is, the number of measures falling within each group. The total of this column is the total number of measures (270), designated by N. This kind of distribution is called a *grouped frequency distribution* since the frequencies of neigh-

Table 13-C
GROUPED FREQUENCY DISTRIBUTION OF THE DATA OF TABLE 13-A

Class Interval	Interval Midpoint	Frequency *f*
225– 375	300	1
375– 525	450	1
525– 675	600	6
675– 825	750	38
825– 975	900	80
975–1125	1050	83
1125–1275	1200	39
1275–1425	1350	17
1425–1575	1500	2
1575–1725	1650	2
1725–1875	1800	0
1875–2025	1950	1
		$\Sigma f = N = 270$

boring individual measures have been gathered together forming a set of frequency groupings.

Each of the equal-sized groups in a grouped frequency distribution is called a *class interval* and is identified by specifying the *class limits* or *class boundaries*. Thus, the class intervals for the frequency distribution of Table 13-C are given in the first column of the table as 225–375, 375–525, etc. The size, or width, of each of the class intervals is designated by i; in the example of Table 13-C, $i = 150$ units (psi).

The degree of condensation achieved by grouping data depends on the number of units included in a class interval—the size of i—since increasing i decreases the number of class intervals necessary to cover the entire distribution. As i increases (and the number of intervals decreases) the overall pattern formed by the data measures tends to emerge more and more clearly in the grouped frequency distribution. This can be seen in Figure 13-1, where three grouped frequency distributions of the data on brick strengths are presented graphically using three different interval widths. The graph on the left is for 24 class intervals and exhibits the greatest irregularity. The figure on the right is the graph of the distribution of Table 13-C, using 12 intervals, and it shows the general pattern of the strength results quite clearly. The center figure shows the same distribution using 18 intervals.

Before we discuss the problem of how many class intervals and what value of i to use, there is a further point about grouping data that should be realized. This is that in the process of grouping, the identity of the individual data item is lost. For example, in the distribution of Table

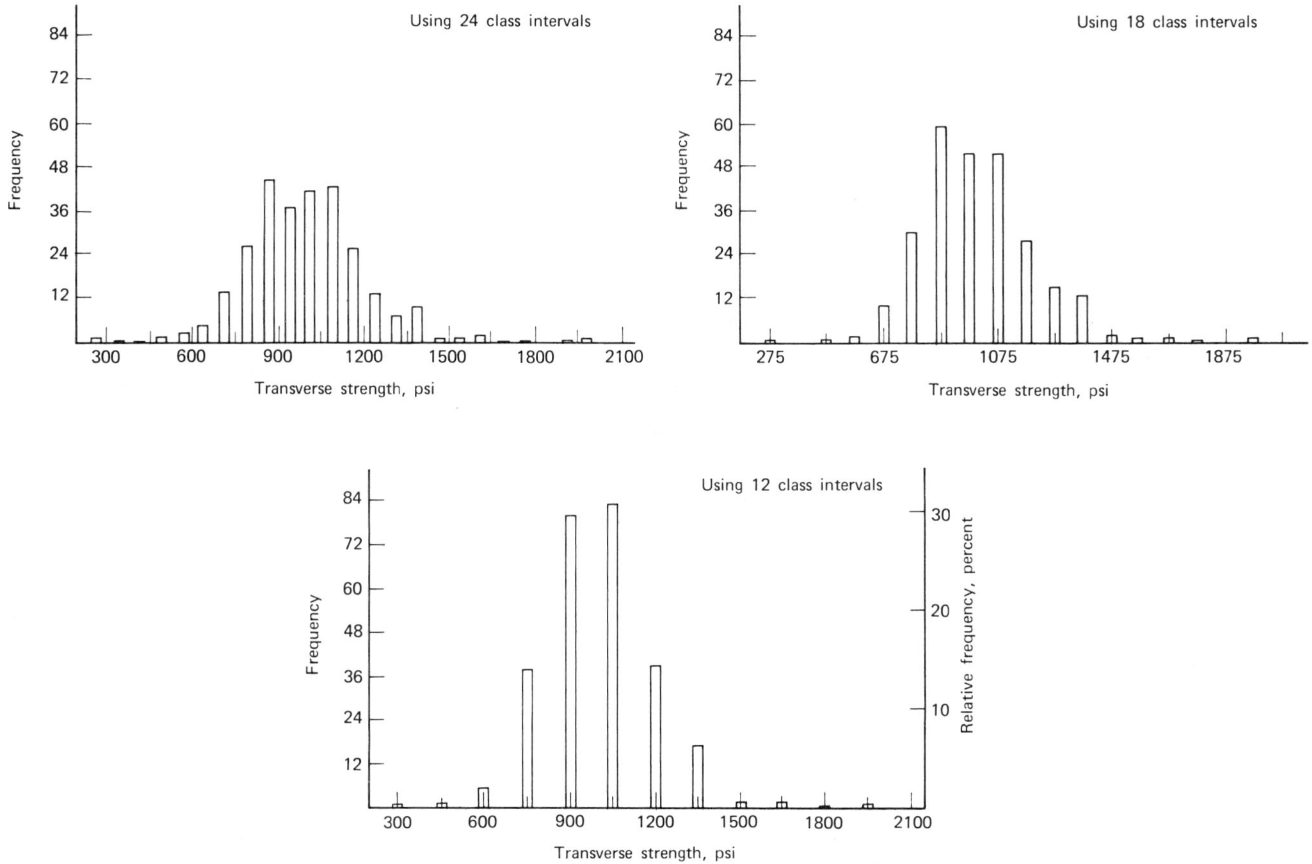

Figure 13–1 Illustrating emergence of group pattern with fewer class intervals (frequency distribution of Table 13-B).

13-C, we know that 38 bricks had strengths between 675 and 825 psi, but we do not know the exact individual strength measures for these bricks once we have put the original data aside and replaced it with a grouped frequency distribution. As i becomes larger, the numerical value of the individual measure becomes more and more obscured. We shall see later that this loss of precision as a result of grouping has a practical implication when certain statistics are computed from grouped frequency distributions.

Number of Class Intervals

Before a grouped frequency distribution can be established from a set of data, the number of class intervals to be employed must be decided. To an extent this is an arbitrary matter, depending on the number of items, the range between the extreme values, and what is to be subsequently done with the grouped distribution. The number of intervals should be few enough to reveal the group pattern, but not so few that precision is seriously impaired through loss of the exact values of the individual measures. A rule of thumb is that from 10 to 20 intervals should be employed, and this usually works quite well for most data. Smaller sets of data will generally require fewer intervals, and when the number of items is as small as 25, say, a grouped frequency distribution of the data will rarely be of much value. A narrow range of measures, in which little variability is exhibited, may preclude grouping too. As we choose the number of intervals, or decide whether measures should be grouped at all, each set of data should be examined individually.

Width of the Class Interval

Once the approximate number of class intervals has been chosen, the width (i) of the class interval required to produce this number of intervals must be determined. A rough estimate of i can be found by dividing the range of the measures (the highest minus the lowest value) by the number of intervals wanted. At this point let us take an example and show how the interval width and the class limits for the grouped frequency distribution might be determined.

Suppose daily samples of 500 test cylinders each were taken from the concrete for a certain job, and these samples were taken over a period of two months, giving about 50 samples. The available data consists of the number of cylinders for each sample not meeting certain minimum strength requirements, and values range from a low of 9 to a high of 61 cylinders. Since N is only about 50, let us assume that we decide on having about 12 class intervals. The approximate interval width would be 4.3 units since

$$\frac{61-9}{12}=4.3$$

Since whole numbers are more convenient than fractions for widths of intervals, especially, as in this case, when the original data are integers, we would choose either 4 or 5 for the value of i.

We shall frequently be using the *midpoints* of class intervals, that is, the central value of any interval (see Table 13-C), and it is much more convenient for these midpoint values to be whole numbers rather than numbers involving fractions. For this reason we generally select class intervals to include an *odd* number of possible measures. This means that if the data are recorded to the nearest whole number, i will be odd. For example, 11 is the midpoint of the interval 10–12 ($i=3$), while 27.5 is the midpoint of the interval 26–29 ($i=4$).

Returning to our example, we would therefore select 5 for the value of i. We might then set up the class intervals shown in Table 13-D, beginning with an interval that includes 9, the lowest value, and terminating with the interval that includes 61, the highest value.

These class intervals of width $i=5$ were chosen to give the convenient midpoint values shown in the center column—10, 15, etc.; but we could have chosen other intervals, such as 5–9, 10–14, etc. (giving midpoint values of 7, 12, etc.) if we had wished. As can be seen, we ended up with 11 intervals.

In this example (Table 13-D), involving numbers of concrete cylinders with below-minimum strengths, the data consists of *whole* numbers only.

Table 13-D
CLASS INTERVALS, MIDPOINTS, AND REAL LIMITS DETERMINED FOR AN EXAMPLE PROBLEM (see text for discussion)

Class Interval	Internal Midpoint	Real Limits
8–12	10	7.5–12.5
13–17	15	12.5–17.5
18–22	20	17.5–22.5
23–27	25	22.5–27.5
28–32	30	27.5–32.5
33–37	35	32.5–37.5
38–42	40	37.5–42.5
43–47	45	42.5–47.5
48–52	50	47.5–52.5
53–57	55	52.5–57.5
58–62	60	57.5–62.5

The class intervals in the left column of the table reflect this fact since they have gaps between intervals: 12–13, 17–18, etc. This brings up the concept of *real limits*, class limits that eliminate such gaps. The real limits for this example are indicated in the right column of the table. Thus, if values such as 12.1 or 12.8 had been possible, they could not have been assigned to either the first or second interval in the left column, whereas recognition of the real limits allows us to assign them properly. Although there are no such problems in this example where only whole numbers are possible, what is usually done is to choose class limits halfway between two possible values of the measure. This avoids gaps between intervals and insures that there be no question about which interval a measurement belongs in.

Let us look at the grouped frequency distribution of brick strengths in Table 13-C and see what class intervals were selected. Here the class limits are given as 225–375, 375–525, etc., and these are real limits chosen midway between two possible measures (the data are recorded to the nearest 10 psi, you will remember). It would also have been acceptable to indicate the class intervals as 230–370, 380–520, etc., but these class limits are not the real limits. Regardless of which method we use for indicating class intervals, the important thing is to be aware of what the real limits are. Each interval of Table 13-C contains 15 possible values, and the intervals have been chosen so that the midpoints are conveniently located at 300, 450, etc. The use of fractions has been completely avoided in this distribution.

The following summarizing points are a general guide for selecting class intervals for a grouped frequency distribution. It is presupposed that there are a sufficient number of measures to warrant grouping.

1. Choose between 10 and 20 intervals, whichever gives the most useful distribution. If necessary, experiment with more than one grouping.
2. Divide the difference between the highest and lowest values of the data by the number of intervals desired to obtain the approximate interval size.
3. Select the interval size so that there are an odd number of possible values in each interval.
4. Position the intervals around convenient midpoint values (e.g., multiples of 5 or 10).

Cumulative Frequency Distribution

Sometimes the total frequency below or above a certain value is of interest. For example, for the distribution of Table 13-E we might wish to

Table 13-E
CUMULATIVE FREQUENCY DISTRIBUTION FOR THE GROUPED FREQUENCY DISTRIBUTION OF TABLE 13-C

Class Interval	Frequency, f	Cumulative Frequency
225– 375	1	1
375– 525	1	2
525– 675	6	8
675– 825	38	46
825– 975	80	126
975–1125	83	209
1125–1275	39	248
1275–1425	17	265
1425–1575	2	267
1575–1725	2	269
1725–1875	0	269
1875–2025	1	270
	$\Sigma f = N = 270$	

know the number of bricks having strengths less than 825 psi, or we might be interested in the number with strengths equal to or greater than 1275 psi. In such cases a *cumulative frequency distribution*, consisting of a series of *cumulative frequencies*, is useful.

The cumulative frequency distribution corresponding to the grouped distribution of brick strengths is given in Table 13-E in the column headed "cumulative frequency." These cumulative frequencies were obtained by a sequence of additions, beginning at the top and proceeding down through the table to the bottom, and each value represents the total frequency for that class interval and all lower intervals of the distribution (all intervals above it in the table). The distribution is derived by first recording the frequency, f, of the first interval, then adding to it the frequency of the second interval, adding to that total the frequency of the third interval, and so on, each new sum representing the cumulative frequency for that and all previous intervals. Thus, in Table 13-E we begin with a value of 1, the frequency of the class interval 225–375, and proceed as follows: 2 $(1+1)$, 8 $(6+2)$, 46 $(38+8)$, etc. The final cumulative frequency, for the highest class interval, will always equal the total frequency, N, since all frequencies lie within or below this interval; in Table 13-E this frequency is 270, the total number of bricks.

Since the cumulative frequency associated with a class interval includes the frequency of that class, the cumulative frequency distribution tells us the total frequency for all values below the upper real limit of each class interval. Hence, the distribution of Table 13-E tells us that 46 bricks have strengths below 825 psi, the upper class limit of the fourth interval. In order to determine the number of bricks *equal to or greater than* 1275 psi, the cumulative frequency corresponding to an upper class limit of 1275 must be subtracted from the total frequency, 270. Using Table 13-E, then, we find that 22 bricks (270–248) have strengths equal to or greater than (i.e., not less than) 1275 psi. We could also have obtained this value by summing frequencies for all class intervals higher than 1275 ($17+2+2+0+1=22$).

PRESENTATION OF FREQUENCY DISTRIBUTIONS

The distributions of Tables 13-B, 13-C, and 13-E are sometimes referred to as *tabular* frequency distributions to distinguish them from graphical presentations, which we shall discuss shortly.

Relative Frequencies

In many cases a frequency distribution can be made more understandable by presenting relative frequencies rather than, or in addition to, actual frequencies. A relative frequency is simply an actual frequency (f) divided by the total number of cases (N), and may be in the form of a percentage or a decimal fraction.

Table 13-F shows the relative frequencies for the grouped frequency

Table 13-F
RELATIVE FREQUENCY FOR THE GROUPED FREQUENCY DISTRIBUTION OF TABLE 13-C

Class Interval	Relative Frequency (Percentage)
225– 375	0.4
375– 525	0.4
525– 675	2.2
675– 825	14.1
825– 975	29.6
975–1125	30.7
1125–1275	14.5
1275–1425	6.3
1425–1575	0.7
1575–1725	0.7
1725–1875	0.0
1875–2025	0.4
	Total = 100.0

$N=270$

distribution of Table 13-C. These frequencies are expressed as percentages and the total must, of course, be 100. Table 13-G contains cumulative relative frequencies, also percentages, corresponding to the cumulative frequency distribution of Table 13-E. In this kind of table, the final cumulative relative frequency is always 100%. Note that in both these relative frequency distributions the total number of measurements is given at the bottom. This number (N) should always accompany a relative frequency distribution when actual frequencies are not also given.

Graphical Representations of Frequency Distributions

All of the frequency distributions discussed in this chapter can be represented graphically. Since most distributions of sufficient size are grouped, however, we will confine ourselves to describing several ways of graphing grouped frequency distributions and will continue using the example of the transverse strengths of 270 bricks.

Figure 13-1 shows graphs of three grouped frequency distributions based on the data of brick strengths. This kind of graph is called a *frequency bar chart,* drawn by centering bars (or lines) on the midpoints of the class intervals, the height of each bar indicating the frequency for that interval. If the frequencies of the various class intervals are marked by a point or tiny circle at the proper height over each midpoint, and these successive points are then connected by a series of straight lines, the resulting graph is called a *frequency polygon.* The frequency polygon

Table 13-G
CUMULATIVE RELATIVE FREQUENCY FOR THE
CUMULATIVE FREQUENCY DISTRIBUTION OF TABLE 13-E

Class Interval	Cumulative Relative Frequency (percentage)
225– 375	0.4
375– 525	0.8
525– 675	3.0
675– 825	17.1
825– 975	46.7
975–1125	77.4
1125–1275	91.9
1275–1425	98.2
1425–1575	98.9
1575–1725	99.6
1725–1875	99.6
1875–2025	100.0
$N=270$	

corresponding to the frequency bar chart in the right of Figure 13-1 (the distribution of Table 13-C) is shown in Figure 13-2. Note that the polygon is completed at either endpoint by dropping a line to the baseline (zero frequency) at the next interval midpoint.

Let us now look more closely at the graphs we have considered so far, the frequency bar chart and frequency polygon. First, there are two axes used: the horizontal baseline, known as the *X*-axis or abscissa, and the vertical so-called *Y*-axis, commonly called the ordinate. The values of the measurements, indicated by the midpoint values of the class intervals, are represented along the horizontal *X*-axis, while the *Y*-axis represents frequency (or sometimes relative frequency). This disposition of the axes should always be followed when graphing a frequency distribution.

Each axis is labeled according to what the axis represents, in this case transverse strength, psi, for the *X*-axis and number of bricks (or frequency) for the *Y*-axis. Adequate labeling of axes should always be done. The lowest values for both axes should start at the intersection of the axes (called the *origin*), at the lower left of the graph. It is not necessary to number the position of every midpoint value or other measure along the baseline, but only certain representative ones at regular intervals. The same policy applies to the frequency scale; as long as enough equally spaced points are identified on the *Y*-axis to make interpretation easy, numbering is adequate. Naturally, every graph should have its own descriptive title, which should be concise but sufficiently informative.

The physical distance selected to represent a unit of measurement or frequency unit is a matter of convenience and convention. Such units should be chosen so that the *Y*-axis is about two-thirds or three-fourths the length of the *X*-axis. Graphs that depart too much from this approximate standard may give a misleading impression at first glance.

A slightly different graphical representation of a grouped frequency distribution is obtained by erecting a rectangular column over the full width of each class interval. The height of each column equals the class frequency. Such a graph is called a *frequency histogram.* The histogram for the grouped distribution of brick strengths is shown in Figure 13-3.

A cumulative frequency distribution can also be represented graphically, and the graphing is carried out in a manner similar to that for the frequency polygon. In this method, however, the points are plotted directly above the upper real limits of each interval rather than over each midpoint. (See Fig. 13-4.) Upper real limits are used because cumulative frequency represents the number of measurements having values below

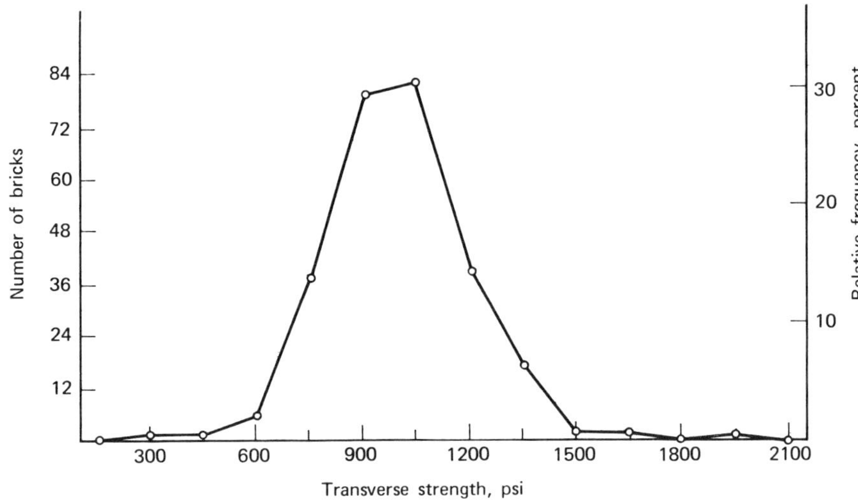

Figure 13–2 Frequency polygon for frequency distribution of Table 13–C.

the value at the upper limit. The height of each point above the *X*-axis is, of course, equal to the appropriate cumulative frequency.

Relative Frequency Scales

The frequency bar chart, frequency polygon and frequency histogram, and the graph of the cumulative frequency distribution, can all be drawn using relative frequencies for the scale of the *Y*-axis. The figures illustrating the distribution of Table 13-C each have a relative frequency scale indicated along the right side. The relative frequencies and cumulative relative frequencies for the distribution are given in Tables 13-F and 13-G, respectively.

Percentiles

A graph of a cumulative frequency distribution with a relative frequency scale of percentages provides a simple means of determining percentiles. A *percentile* is the value of the variate (in our example, a brick strength) below which a given percentage of the total number of cases lie. Thus, the 30th percentile is the value below which 30% of the measurements lie, and the 65th percentile the value below which 65% lie. The concept of percentiles is a simple one, but quoting certain select percentiles can be quite helpful when describing a distribution. The 50th percentile, also

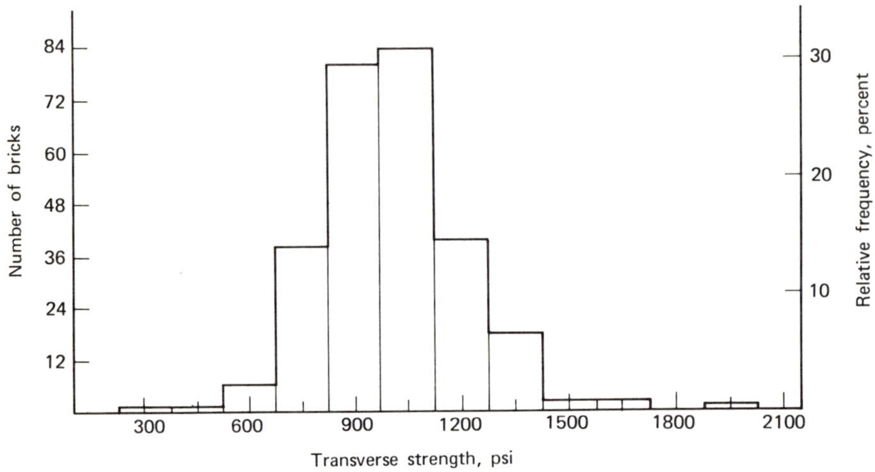

Figure 13–3 Frequency histogram for frequency distribution of Table 13–C.

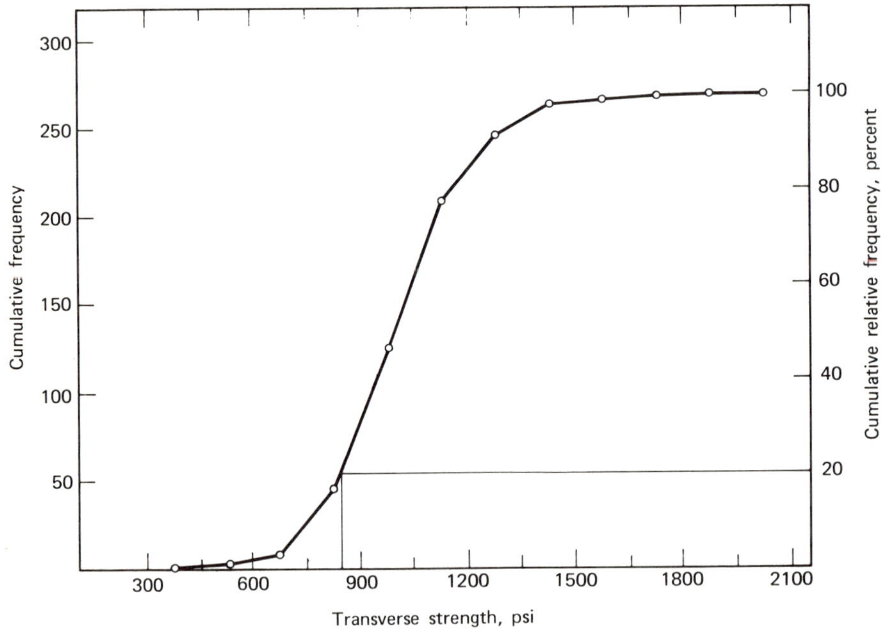

Figure 13–4 Graphical presentation of a cumulative frequency distribution showing determination of the 20th percentile (distribution of Table 13–E).

called the median, is particularly important, as we shall see in the next chapter.

Although percentiles can be computed from a tabular frequency distribution, the simplest method of determining any percentile is by inspection of the graph of the cumulative frequency distribution as mentioned above. For example, let us see how the 20th percentile can be read from the graph of Figure 13-4. To do this we have to draw a horizontal line from the 20% point on the relative frequency scale across to the curve, and then drop a vertical line down to the X-axis from this point on the curve. These auxiliary lines for this example are shown in the figure, and the value we read on the baseline tells us the 20th percentile, in this case about 840 psi.

Form of a Frequency Distribution

Grouped frequency distributions, and hence their graphical representations, occur in a number of characteristic shapes or forms. Since form generally provides important information about a distribution, it would be worthwhile to look at some of these types. For convenience we will illustrate this using frequency polygons.

Figure 13-5 shows a variety of *symmetrical* distributions. Such distributions are symmetrical with respect to a center line perpendicular to the baseline because if the polygon is folded in half along this line, the two sides are identical in shape.

The distributions in Figure 13-5*a* are examples of symmetrical bell-shaped distributions, and a great many phenomena take this general shape. The middle curve in *a* approximates the *normal distribution*, a specific form of a symmetrical bell-shaped distribution. The normal distribution is of considerable importance in statistics and will be discussed in some detail in Chapter 15. A useful characteristic for comparing bell-shaped distributions is the degree of peakedness of the graph relative to the normal curve, technically known as *kurtosis*. A curve more peaked than the normal distribution, such as the taller curve in Figure 13-5*a*, is called *leptokurtic*, while the flatter curve in the figure is said to be *platykurtic*. Distribution *b* is a *rectangular distribution*, the frequencies of each of the class intervals being the same. Distribution *c* is a symmetrical distribution intermediate in form between the rectangular and bell-shaped distributions.

Three distributions that are not symmetrical are shown in Figure 13-6. These are illustrations of skewed freqency distributions. Any asymmetrical distribution in which the measures "tail off" more in one direction

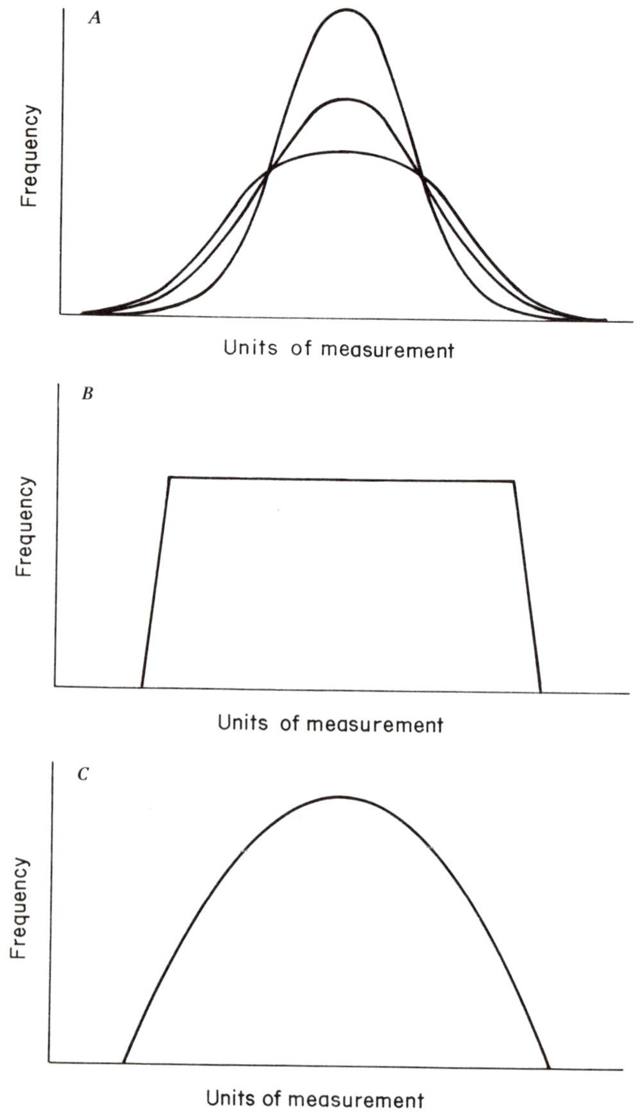

Figure 13–5 Illustrations of symmetrical frequency distributions.

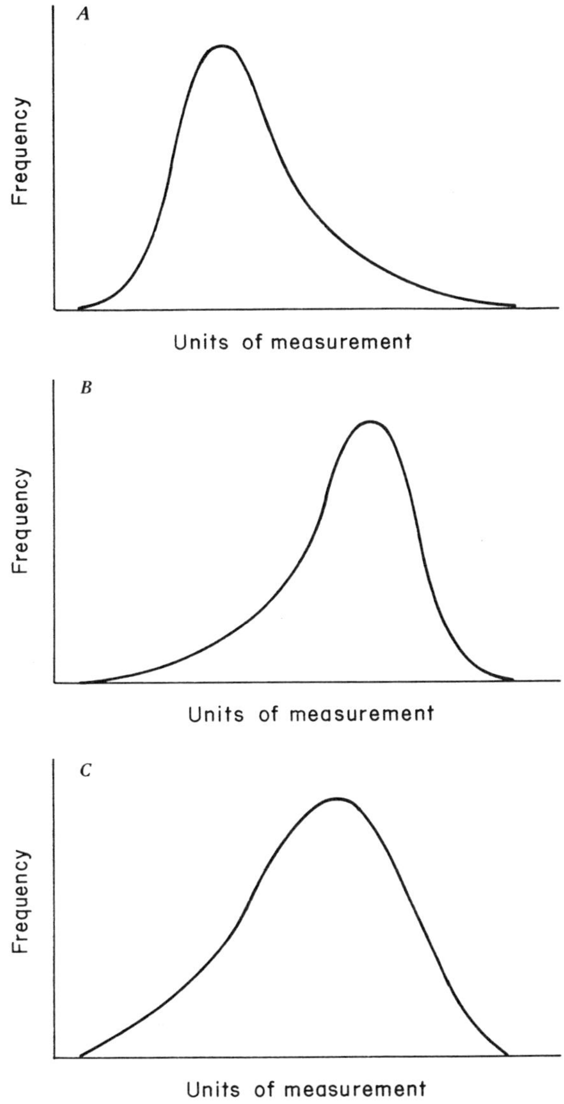

Figure 13–6 Illustrations of skewed frequency distributions.

than the other is said to be *skewed*, and this property of a distribution is called *skewness*. Degree of skewness refers to the extent that a distribution departs from symmetry. Thus distribution *c* in Figure 13-6 is less skewed than *b* or *a*.

An important distinction for skewed distributions is the direction of skewness, determined by which end of the baseline scale the measures tend to trail off at. If the tail extends out toward the upper end of the scale, the distribution is said to be *positively skewed* or skewed to the right. Distribution *a* of Figure 13-6 is positively skewed. Distribution *b*, with the tail at the lower end, is said to be *negatively skewed* or skewed to the left.

Skewness in a distribution can arise from natural or from artificial causes. As an example of the latter, consider the testing of concrete specimens. If the obviously poor or defective specimens are discarded and not tested, the distribution of strengths, for example, of the remaining specimens will be positively skewed due to the selective elimination of specimens having presumably the lowest strengths. On the other hand, if we measured the widths of all cracks in the surface of a test section of pavement, we would naturally expect the distribution of these crack widths to be skewed to the left. This is because the frequency of occurrence of very large cracks is much smaller than that of tiny cracks.

CHAPTER 14

CHARACTERISTICS OF FREQUENCY DISTRIBUTIONS

MEASURES OF CENTRAL TENDENCY

The distributions in Figures 13-2 and 13-6 are similar in that each has a more or less central region where the measures tend to be concentrated and upper and lower "tails" where measures are relatively infrequent.

If we are looking for a single value of the variate to represent or typify the entire distribution, it would surely lie within this range where the values tend to pile up. This measure of central tendency, used to represent or characterize a group of data as a whole, is an example of a *descriptive statistic*. Such statistics are useful in conveying information about a set of data in a relatively simple form, and we shall discuss a number of descriptive statistics in this chapter.

Several measures of central tendency have been devised, but we will only discuss the three most common: the mean, median, and mode. Choice of which measures to use in a particular instance depends on the nature of the distribution and on the purposes for which the data have been collected. More will be said about this after discussing these statistics.

The Mean

The mean, or arithmetic mean, is what is commonly known as the average and is found by the familiar process of summing the values and dividing this total by the number of values being averaged.

The mean can be thought of as a point of balance, or a fulcrum, in the sense that the values on one side of the mean deviate from it to a total extent equal to the deviations on the other side. Thus, if the deviations of values above the mean are considered positive (+) and deviations of values below the mean are considered negative (−), the algebraic sum (taking the signs into consideration) of all deviations will be zero. This fact is useful when checking the accuracy of a computed mean and also constitutes the basis of a method for computing the mean. We will be working with deviations from the mean later in this chapter, in the section on variability, and there the concept will again be useful in checking the correctness of computed deviations.

Calculating the Mean. The mean of a set of values of the variate X we shall call $\overline{X}$ (X-bar). We have already mentioned the familiar method for computing X, and will now give an example. The formula for computing the mean of a set of values X is

$$X=\frac{\Sigma X}{N}$$

where N is the number of values and Σ (Greek capital letter *sigma*) means to sum all values of the variable following it—X in this case. Table 14-A shows batch weights (pound) of cement for 10 successive 1-cu-yd samples of concrete together with the computation of the mean weight.

When N is large, it is sometimes simpler to compute the mean by summing deviations from an assumed mean rather than summing the X values, since these deviations are usually much smaller numbers. This method is based on the fact, as we saw above, that the deviations of measures from their mean algebraically sum to zero. To show how this method works, let us take the example of cement weights and show how the mean can be calculated by first assuming or guessing a fictitious mean.

The formula for this "assumed-mean" method is

$$\overline{X}=\overline{X}'+\Sigma d'/N$$

where $\overline{X}'$ is an assumed mean and d' are the deviations of the values from this assumed mean; $\Sigma d'$ means to sum (algebraically) all such deviations.

Table 14-A
CALCULATION OF X FROM UNGROUPED DATA

Pounds of Cement X
587
592
589
582
589
586
601
590
595
580
$\Sigma X = 5891$
$\overline{X} = \frac{\Sigma X}{N} = \frac{5891}{10} = 589.1$

In Table 14-B we have assumed the mean to be 590 and figured the individual deviations from this mean. We could just as well have chosen some other convenient number such as 585 or 580 (the latter avoiding negative numbers); the process would be the same. At the bottom of the table, the assumed mean is adjusted by an amount equal to the mean of these deviations to give the true mean.

Table 14-B
CALCULATION OF $\overline{X}$ FROM UNGROUPED DATA USING AN ASSUMED MEAN OF 590 (Same Data as in Table 14-A)

X	Deviation from Assumed Mean d'
587	− 3
592	+ 2
589	− 1
582	− 8
589	− 1
586	− 4
601	+11
590	0
595	+ 5
580	−10
	$\Sigma d' = -9$

$$\overline{X} = \overline{X}' + \frac{\Sigma d'}{N} = 590 + \frac{-9}{10} = 590 - 0.9 = 589.1$$

Grouped Distributions. When the number of measures (N) is at least fairly large, it is usually better to work with a grouped frequency distribution rather than manipulating a large quantity of individual data. We discussed grouped frequency distributions in Chapter 13, and here it is necessary to show how the basic method of calculating a mean must be modified, since we are working with condensed data rather than the original measures.

The mean of a grouped frequency distribution can be expressed as

$$\overline{X} = \frac{\Sigma f(MP)}{N}$$

a formula comparable to the familiar one given above for ungrouped data. In this expression, f is the frequency of a class interval, MP its midpoint value, and Σ tells us to sum the products, $f \times MP$, for all intervals before dividing by N, the total frequency or total number of values. In other words, the formula says to form a product for each class interval of the grouped frequency distribution equal to the frequency of the interval times its midpoint value, and then to sum these *products*; this sum is then divided by N to arrive at the mean (X).

Table 14-C shows the class intervals and midpoint values together with the frequencies for the data on brick strengths from Chapter 13. The

Table 14-C
CALCULATION OF $\overline{X}$ FROM A GROUPED FREQUENCY DISTRIBUTION
(Data from Table 13-C)

Class Interval	Class Midpoint MP	Frequency f	$f(MP)$
225 375	300	1	300
375– 525	450	1	450
525– 675	600	6	3,600
675– 825	750	38	28,500
825– 975	900	80	72,000
975–1125	1050	83	87,150
1125–1275	1200	39	46,800
1275–1425	1350	17	22,950
1425–1575	1500	2	3,000
1575–1725	1650	2	3,300
1725–1875	1800	0	0
1875–2025	1950	1	1,950
		$N = \Sigma f = 270$	$\Sigma f(MP) = 270{,}000$

$$\overline{X} = \frac{\Sigma f(MP)}{N} = \frac{270{,}000}{270} = 1000$$

f times MP products are then figured for each line and summed. The final computation for X is shown at the bottom of the table. This kind of format, or a comparable one, should always be used when calculating $\overline{X}$ for a grouped frequency distribution in order to minimize the possibility of error and make the computation work as easy as possible.

The process of summing the products of frequencies times interval midpoints over all intervals of a grouped frequency distribution is comparable to adding all the data items when computing the mean for simple, ungrouped data. It should be said, however, that a small error is introduced by using midpoint values of intervals. This usage assumes that the frequencies within each interval are distributed *uniformly* throughout the interval, so that they can be thought of as being concentrated at the midpoint. Actually, for a particular interval, this will probably not be exactly the case, but over the entire distribution these slight discrepancies will tend to balance each other since in some intervals frequencies will average above the midpoint while in others the reverse will be true. Thus, a mean calculated from a frequency distribution of grouped data will not generally be the exact mean of the original data, but usually the difference is quite small and can be ignored. For example, if X is calculated for the 270 brick strengths of the original ungrouped data (Table 13-A), the value obtained is 999.8 compared to the value of 1000.0 obtained in Table 14-C for the grouped frequency distribution.

The multiplications called for when working with a grouped distribution can be made much easier by assuming the mean to be at the midpoint of some interval and summing products of frequencies times scaled midpoint deviations (deviations in terms of the number of intervals) rather than multiplying by the midpoint values themselves. This method is comparable to the "assumed-mean" method for ungrouped data discussed previously. The "assumed-mean" formula for a grouped distribution is

$$\overline{X}=\overline{X'}+\frac{i\Sigma fd'}{N}$$

in which

X' = an assumed mean at the midpoint of some interval
i = the uniform interval size
f = the frequency of an interval
d' = the deviation of an interval from the assumed mean, expressed as the number of intervals above or below (+ or −) this "mean"

$\Sigma fd'$ means to sum the products $f \times d'$ for all intervals
N = total frequency

To show how this method works, consider again the distribution of Table 14-C. Let us assume, for this example, that the mean is the midpoint value of the class interval with the highest frequency; thus, $\bar{X}' = 1050$ (see Table 14-D). The deviations of the midpoints of the several intervals from $\bar{X}'$ are indicated in the column headed d'. These deviations are multiplied by the corresponding values of f in each line of the table, and these products are then summed. This sum must then be multiplied by the interval width (i), which is 150. As can be seen, completing the calculation gives a value of $\bar{X}$ of 1000—the same value obtained before when we used the midpoint values directly. In contrast to the method of Table 14-C, all the multiplications in the right-hand column of Table 14-D are simple enough to be done mentally.

The Median

Another common measure of central tendency is the median (Mdn.). The median of a distribution is simply the central or middle value (or the

Table 14-D
CALCULATION OF $\bar{X}$ FROM A GROUPED FREQUENCY DISTRIBUTION USING AN ASSUMED MEAN OF 1050
(Same Data as Table 14-C)

Class Interval Midpoint	Deviation from Assumed Mean in terms of Number of Classes d'	Frequency f	fd'
300	−5	1	− 5
450	−4	1	− 4
600	−3	6	−18
750	−2	38	−76
900	−1	80	−80
1050	0	83	0
1200	1	39	39
1350	2	17	34
1500	3	2	6
1650	4	2	8
1800	5	0	0
1950	6	1	6
		$N = \Sigma f = 270$	$\Sigma fd' = -90$

$$\bar{X} = X' + \frac{i\Sigma fd'}{N} = 1050 + \frac{150(-90)}{270} = 1050 - \frac{150\ (1)^*}{3} = 1050 - 50 = 1000$$

* Cancelling 90 into 270.

mean of the two central values if there are an even number of measures) when the values are arranged in order of size. Thus, the median is simply a name for the 50th percentile—the point in a distribution above which 50% of the values lie and below which 50% lie.

The procedure for determining the median of a set of ungrouped data is the obvious one of first arranging the items in order of size and then taking the middle value if there are an odd number of items or the point midway between the two central values if there are an even number. Thus, the median of the numbers 2, 5, 8, 14, 19 is 8, while the median of the numbers 2, 5, 8, 14, 19, 36 is 11 (midway between 8 and 14).

Grouped Distributions. The process of determining the median for a grouped distribution is to locate in which interval the central value (or central pair of values) lies and then interpolating within that interval to find the value. This is illustrated in the following example for the distribution of Table 14-E.

Since there is a total of 25 items, the median is the value of the 13th item when they are ranked. The three lowest intervals include a total of 11 items (1+3+7), and the fourth interval contains another 8. Clearly, the 13th item must lie within this fourth interval, the interval 25–29, and we only have to interpolate correctly to obtain the median. In this case the 13th value is the second one (next to the lowest) in this fourth interval since 11 values fall below the interval. Since the interval 25–29 has a frequency of 8 items, the median value is one-fourth of the way into it ($2/8 = 1/4$). In this example the interval width is 5, and the median is therefore $1/4 \times 5$, or 1.25, units above the lower limit of this interval. Thus the median is $24.5 + 1.25 = 25.75$. Note that the *real* lower limit (24.5) has

Table 14-E
GROUPED FREQUENCY DISTRIBUTION FOR ILLUSTRATING DETERMINATION OF THE MEDIAN (see text)

Class Interval	Real Class Limits	Frequency f
10–14	9.5–14.5	1
15–19	14.5–19.5	3
20–24	19.5–24.5	7
25–29	24.5–29.5	8
30–34	29.5–34.5	4
35–39	34.5–39.5	2
		$N=25$

been used. This is necessary since the interval really includes values down to and including this value.

As we saw in computing the mean for grouped data, it was necessary to assume a uniform distribution of frequencies within each entire interval. This assumption has also been made in the above interpolation for obtaining the median, and therefore the value obtained is not necessarily identical to the median determined if the original data items had been used. The 13th value of the ungrouped frequency distribution may have been anywhere from 24.5 up to 29.5. The grouped distribution only tells us that it lies within this interval. If we don't have access to an ungrouped distribution of the data, the best we can do is assume that frequencies are distributed uniformly within an interval and interpolate accordingly.

The Mode

The mode is the final measure of central tendency we shall take up here and is conceptually the simplest. For ungrouped data, the mode (*Mo.*) is defined as that value which occurs most frequently. For data grouped into class intervals, the mode is the midpoint of the interval with the greatest frequency. In either case, the mode can be determined as the peak of the graphed frequency distribution. If the distribution has not been plotted, the mode can be easily determined by glancing over the ungrouped frequency distribution and noting the value or interval of values that is most frequent. For the grouped distribution of Table 14-C, the mode is 1050.

Occasionally, a frequency distribution will have two humps or peaks in its graph indicating two points of maximum frequency. This can occur when the data is not homogeneous and contains values from two different distributions. Such a distribution is called *bimodal* and has two modes.

Relations Between Measures of Central Tendency

Let us now look at the relations between the three measures of central tendency we have discussed—*X*, *Mdn.* and *Mo.*

In a symmetrical, bell-shaped distribution, such as was discussed in Chapter 13, the three measures will coincide. This is because the point of balance and the value dividing the frequencies into two equal parts both lie at the peak or mode of the distribution. (See Fig. 14-1.)

If the distribution is skewed (distorted from symmetry), however, $\overline{X}$, *Mdn.*, and *Mo.* will fall at different points on the curve. A positively skewed distribution, you will remember, is one in which there are a few relatively extreme values at the high end of the scale that aren't matched

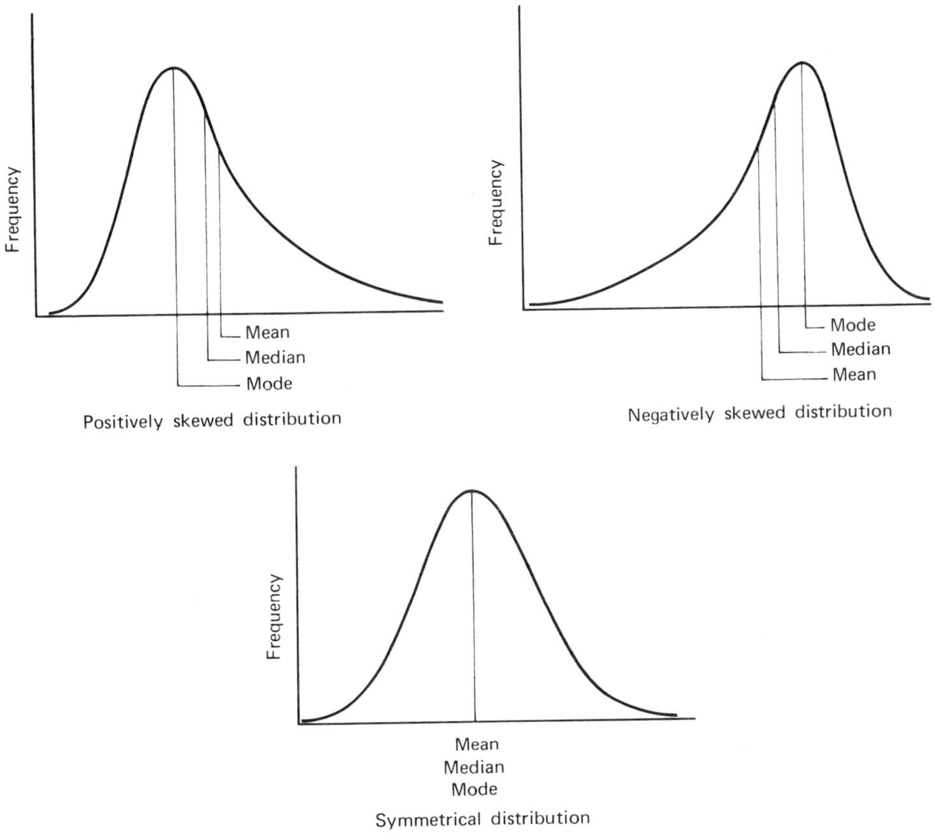

Figure 14–1 Relative positions of mean, median, and mode for skewed and symmetrical frequency distributions.

by corresponding low values. These values cause the tail in the upper end of the distribution, and they have an effect of "pulling" $\overline{X}$ and *Mdn.* above the mode. Furthermore, $\overline{X}$ will be greater than the *Mdn.* since the high *values* of these extreme items affect the mean, whereas only the *number* of such items acts to increase the *Mdn.* above the *Mo.*

If the same reasoning is applied to a negatively skewed distribution (one in which the tail is at the low end of the distribution), it can be readily seen that $\overline{X}$ and *Mdn.* will both lie *below* *Mo.* with $\overline{X}$ closer to the tail. Thus the relative position of the three measures of central tendency for a skewed frequency distribution will always be mode, median, mean when going from the peak to the tail. As we have said, these measures have the same value in a symmetrical distribution. These relationships are illustrated in Figure 14-1.

Measure of Skewness. The fact that the mean is displaced from the mode in a skewed distribution gives us a convenient measure of skewness (degree of distortion from symmetry). This measure is: $(\overline{X} - Mo.)/SD$. The term in the denominator is the standard deviation, which we will take up in the next section, and dividing by this statistic allows comparison of skewness between distributions.

MEASURES OF DISPERSION

We have seen in the previous section how distributions can be described using measures of central tendency. Another very important characteristic of frequency distributions is the disperson or variability of the measures. For example, suppose the mean compressive strength of a sample of concrete test cylinders is 3000 psi. If, in order to satisfy design requirements, concrete strengths must be at least, say, 2500 psi, we might be interested in the proportion of cylinders with strengths below this level. We would then want some knowledge of the scatter of the test results, information not conveyed by knowing the mean alone. Test strengths may, for example, be distributed between 2800 and 3200 psi or they may range from 2000 to 4000 psi. (See Fig. 14-2.) In both these hypothetical cases the mean is 3000 psi, but the dispersion of values differs markedly between them; only in the second case would any cylinders have strengths below the minimum requirement of 2500 psi.

There are several measures of dispersion for frequency distributions, the most useful being the standard deviation. Before taking up the standard deviation, however, we will discuss two conceptually simpler measures of variability.

The Range

The range (R) is the simplest such measure and provides a common-sense indication of variability. It is simply the distance between the lowest and highest values of a distribution. The ranges of the distributions in Figure 14-2 are 400 psi (3200–2800) for the narrower distribution and 2000 psi (4000–2000) for the other.

The range is only a rough index of variability since its value depends only on the two extreme measures in a set of data. These extreme values

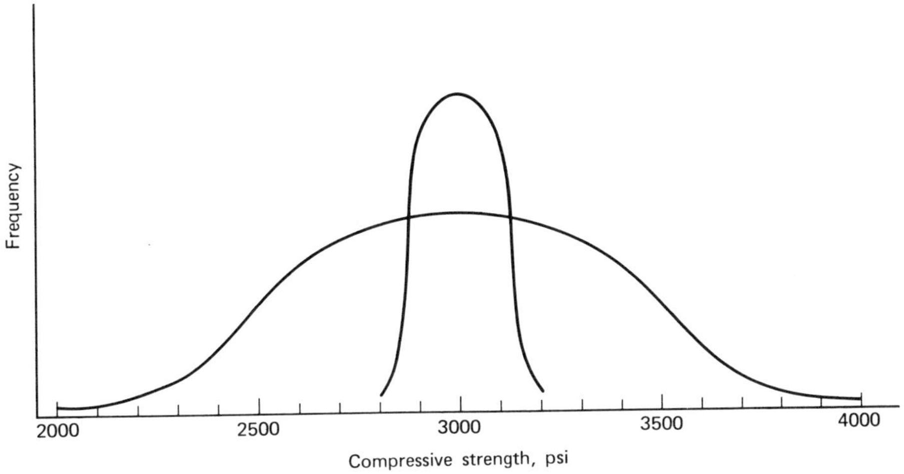

Figure 14–2 Two hypothetical distributions have the same mean but different variabilities (see text).

may be capricious, and R will reflect this. For example, if through faulty test procedures one of the cylinders from the narrower distribution in Figure 14-2 was cast with only 2000-psi compressive strength, that distribution would have a range of 1200 rather than 400 psi—a tripling of R due to only one case. The range has the additional weakness of not indicating anything about the dispersion of the remaining data items, as other variability measures do.

Mean Deviation

In discussing the mean, we stated that the algebraic sum of the deviations of measures from the mean of those measures is zero. This means that the sum of all deviations above the mean exactly equals the sum of deviations below. In this section, and in the next section on standard deviation, we will see how these deviations constitute a convenient basis for measures of dispersion.

Examine the two frequency distributions of Table 14-F. Both distributions have the same $\overline{X}$, 25, but they obviously differ considerably in variability, as can be seen from the ranges. For each distribution the deviations are given in the column headed x and are equal to $X - X$ (each values minus the mean). As expected, these deviations sum to zero for each of the two distributions.

Table 14-F
CALCULATION OF THE MEAN DEVIATION FOR TWO DISTRIBUTIONS WITH THE SAME MEAN AND DIFFERENT VARIABILITIES

X	Deviation from $\overline{X}$ x	X	Deviation from $\overline{X}$ x
19	−6	11	−14
20	−5	15	−10
22	−3	18	−7
22	−3	20	−5
23	−2	21	−4
24	−1	23	−2
24	−1	23	−2
25	0	24	−1
26	+1	26	+1
26	+1	28	+3
26	+1	29	+4
28	+3	31	+6
29	+4	32	+7
29	+4	34	+9
32	+7	40	+15
$\Sigma X=375$	$\Sigma x=0$ $\Sigma x=42$	$\Sigma X=375$	$\Sigma x=0$ $\Sigma x=90$

$$\overline{X}=\frac{\Sigma X}{N}=\frac{375}{15}=25$$

Mean deviation $=\frac{\Sigma x}{N}=\frac{42}{15}=2.8$

Mean deviation $=\frac{\Sigma x}{N}=\frac{06}{15}=6.0$

If, however, we disregard the signs of the deviations and sum them (indicated by $\Sigma 1x1$), it can be seen that the total is much greater for the distribution with greater variability (90 as opposed to 42). It can be readily seen, then, that the total of the deviations of a distribution *without regard to their signs*—the total of the *absolute values* of the deviations—directly reflects the variability of the distribution. Since this total also depends on the number of items contributing deviations to it, the sum is divided by N. This gives a mean, or average, deviation, and permits comparison of variability between distributions of different sizes. The average deviations of the two distributions are shown at the bottom of the table.

A few additional steps beyond the mean deviation lead to the standard deviation.

Standard Deviation

In computing the mean deviation, we summed the absolute values of the deviations from the mean and divided by N to obtain an average deviation. A measure of much greater usefulness, particularly, as we shall see in the next chapter, when applied to the normal distribution, is the standard deviation (SD). In computing this statistic, we first square each deviation (a step that makes all signs positive), then sum these squared deviations, divide this sum by N to get a mean square (squared) deviation, and finally take the square root of this mean square deviation to arrive at a statistic in the same units as the data (since squaring produced squared units). This process is indicated in the following formula:

$$SD = \sqrt{\frac{\Sigma x^2}{N}}$$

where

SD = standard deviation
x = the individual deviations from the computed mean
Σ = "summation of"—in this case the squared deviations
N = the number of values

Since each deviation, x, is figured by taking $X-X$ (each value minus the mean), the formula can be written

$$SD = \sqrt{\frac{\Sigma(X-\overline{X})^2}{N}}$$

where $\Sigma(X-\overline{X})^2$ means that $(X-\overline{X})^2$ is figured for each value of X (for each data item) and then summed over all values.

The process can be summarized in five steps:

1. Find $\overline{X}$ (by summing the values and dividing by N).
2. Subtract $\overline{X}$ from each X to obtain deviations, x. (As a check at this point, the algebraic sum of these deviations must be zero or there has been a mistake in subtraction or in computing $\overline{X}$.)
3. Square each x.
4. Sum the x^2, then divide by N.
5. Take the square root of the result to get SD.

The example in Table 14-G illustrates the entire process for the data in the left side of Table 14-F.

The result of step 4, above ($\Sigma x^2/N$), is called the *variance* and is equal to SD^2. The variance is a statistic important for certain more advanced

Table 14-G
CALCULATION OF STANDARD DEVIATION BY BASIC METHOD

X	Deviation from X x	Squared Deviation x^2
19	−6	36
20	−5	25
22	−3	9
22	−3	9
23	−2	4
24	−1	1
24	−1	1
25	0	0
26	1	1
26	1	1
26	1	1
28	3	9
29	4	16
29	4	16
32	7	49
$\Sigma X = 375$	$\Sigma x = 0$	$\Sigma x^2 = 178$

$$X = \frac{\Sigma X}{N} = \frac{375}{15} = 25$$

$$SD = \sqrt{\Sigma x^2/N} = \sqrt{\frac{178}{15}} = \sqrt{11.87} = 3.45$$

statistical techniques, and we will not be discussing it further. Its units of measurement are the square of the units of the data, and its square root, as we have seen, is *SD*.

For computational purposes, there is a formula for *SD* that is usually simpler than the basic formula, since it eliminates the need for working with deviations altogether. This formula is the algebraic equivalent of the above formula in that the two can be shown to be identical through algebraic manipulation. Even though the basic formula may not be used in computing *SD*, it is important to understand and remember it since it clearly shows what *SD* is—the root of the mean square deviation. The formula given below is for greater facility in computation; it is *not* evident from it what *SD* actually is. The formula is

$$SD = \sqrt{\frac{\Sigma X^2 - X^2}{N}}$$

This formula tells us to square the individual data values (not the deviations), sum these squares, and divide the total by *N*, and then, before taking the square root, to subtract the square of the mean from this result. This gives the variance (equivalent to $\Sigma x^2/N$ in the basic formula), and taking its square root is the final step in computing *SD*. The process for calculating the standard deviation by this method can also be summarized in five steps:

1. Square each data measure, X.
2. Sum the X^2, then divide by N.
3. Find $\overline{X}$ and $\overline{X}^2$.
4. Subtract X^2 from the sum X^2 divided by N (the result of step 2). ($\frac{\Sigma X^2}{N}$ will always be larger than X^2.)
5. Take the square root of the result to get *SD*.

To show that this procedure will give the same result for *SD* as use of the basic formula will, *SD* is computed in Table 14-H according to the above method for the data used before (Table 14-G). As can be seen,

Table 14-H
CALCULATION OF STANDARD DEVIATION WITHOUT USING DEVIATIONS (same data as Table 14-G)

X	X^2
19	361
20	400
22	484
22	484
23	529
24	576
24	576
25	625
26	676
26	676
26	676
28	784
29	841
29	841
32	1024
$\Sigma X = 375$	$\Sigma X^2 = 9553$

$$\overline{X} = \frac{\Sigma X}{N} = \frac{375}{15} = 25$$

$$SD = \sqrt{\Sigma X^2/N - X^2} = \sqrt{\frac{9553}{15} - 25^2} =$$

$$\sqrt{636.87 - 625} = \sqrt{11.87} = 3.45$$

both methods give a value of 3.45 for *SD*, but this simpler procedure involves one less column since deviations are not needed. A table of squares should be used with this method.

Grouped Distributions. If we are calculating *SD* for a grouped frequency distribution, the process just discussed must be slightly modified. We must first square the midpoint value of each class interval and multiply each of these squares by the corresponding frequency of the interval. These products are then summed and divided by *N*; the remaining steps are the same as for ungrouped data. The formula for calculating *SD* for grouped data is

$$SD = \sqrt{\frac{\Sigma fX^2 - \overline{X}^2}{N}}$$

where X in this case is the midpoint value of an interval and f its frequency; ΣfX^2 means to sum all the individual products fX^2. This procedure is illustrated by the following example:

The modulus of rupture (in psi) was determined on 144 concrete test beams. The frequency distribution is shown in Table 14-I and we wish to calculate *SD* for this distribution. The necessary columns and arithmetic are shown in the table. Note that the column fX is included in order to enable us to calculate X. A table of squares is extremely helpful in this and similar problems of computing standard deviations.

Scaling Down for Convenience in Computing SD. When data values (and hence midpoint values of intervals) are so large that squaring them produces awkwardly large numbers, a useful device is to scale them down. This is done by subtracting a single convenient-sized constant from each measure. This is equivalent to moving the entire distribution down the scale (to the left on the X-axis) by the amount of the constant and does not change the *shape* of the distribution nor, therefore, its variability or *SD*—although it does, of course, change X. This can be seen in Figure 14-3 where the distribution of Table 14-I is shown together with the same distribution shifted down the X-axis by an amount equal to 500 psi. It should be clear from Figure 14-3 that the variability of the items after shifting the distribution down remains unchanged.

If we subtract 500 psi (a convenient round number) from the midpoint values of the class intervals in Table 14-I, we get the values in the

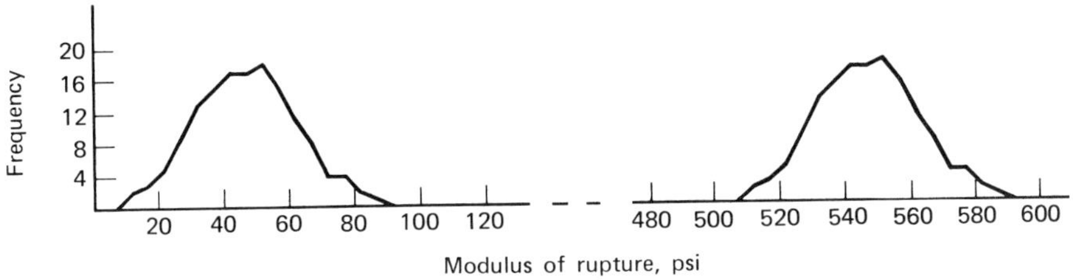

Figure 14-3 Moving a distribution down the X-axis scale to facilitate the calculation of SD (see text).

Table 14-I
CALCULATION OF STANDARD DEVIATION FOR GROUPED FREQUENCY DISTRIBUTION

Class Interval	Interval Midpoint X	Frequency f	fX	X^2	fX^2
510–514	512	2	1,024	262,144	524,288
515–519	517	3	1,551	267,289	801,867
520–524	522	5	2,610	272,484	1,362,420
225–529	527	9	4,743	277,729	2,499,561
530–534	532	13	6,916	283,024	3,679,312
535–539	537	15	8,055	288,369	4,325.535
540–544	542	17	9,214	293,764	4,993,988
545–549	547	17	9,299	299.209	5,086.553
550–554	552	18	9,936	304,704	5,484,672
555–559	557	15	8,355	310,249	4 653,735
560–564	562	11	6,182	315,844	3,474,284
565–569	567	8	4,536	321,489	2,571.912
570–574	572	4	2,288	327,184	1,308,736
575–579	577	4	2,308	332,929	1,331,716
580–584	582	2	1,164	338,724	677,448
585–589	587	1	587	344,569	344,569
		$\Sigma f = N = 144$	$\Sigma fX = 78{,}768$		$\Sigma fX^2 = 43{,}120{,}596$

$$\overline{X} = \frac{\Sigma fX}{N} = \frac{78{,}768}{144} = 547$$

$$SD = \sqrt{\frac{\Sigma fX^2}{N} - X^2} = \sqrt{\frac{43{,}120{,}596}{144} - 547^2} =$$
$$\sqrt{299{,}448.58 - 299{,}209} = \sqrt{239.58} = 15.5$$

Table 14-J
CALCULATION OF STANDARD DEVIATION FROM SCALED-DOWN DATA
(distribution from Table 14-I)

Class Interval Midpoint X	Scaled X $X' = X - 500$	f	fX'	$(X')^2$	$f(X')^2$
512	12	2	24	144	288
517	17	3	51	289	867
522	22	5	110	484	2,420
527	27	9	243	729	6,561
532	32	13	416	1,024	13,312
537	37	15	555	1,369	20,535
542	42	17	714	1,764	29,988
547	47	17	799	2,209	37,553
552	52	18	936	2,704	48,672
557	57	15	855	3,249	48,735
562	62	11	682	3,844	42,284
567	67	8	536	4,489	35,912
572	72	4	288	5,184	20,736
577	77	4	308	5,929	23,716
582	82	2	164	6,724	13,448
587	87	1	87	7,569	7,569
		$\Sigma f = N = 144$	$\Sigma fX' = 6{,}768$		$\Sigma f(X')^2 = 352{,}596$

$$\overline{X'} = \frac{\Sigma f(X')}{N} = \frac{6768}{144} = 47$$

$$SD = \sqrt{\frac{\Sigma f(X')^2}{N} - (X)^2} = \sqrt{\frac{352{,}596}{144} - 47^2} = \sqrt{2448.58 - 2209} = \sqrt{239.58} = 15.5$$

second column of Table 14-J. These values are much smaller and their squares more easily manageable. The calculation of SD is then carried out on these reduced values, as shown in Table 14-J, and the final value of SD (15.5) is the same as was obtained in Table 14-I.

Check on Gross Errors. Since the computation of SD, by whatever method, is rather involved and there is considerable opportunity for arithmetic errors, it would be extremely useful to have some simple means of checking the general accuracy of results. Such a check can be made by comparing the size of SD with R (range), whose value is easy to obtain. For a distribution with a large N, R will generally be *approximately* six times SD. For smaller distributions (N less than about 100) the ratio of R to SD will be somewhat smaller, probably about 4 or 5 to 1; with very small sets of data (N less than 20, say) the ratio may be as low as 3 to 1.

It should be emphasized that these relationships are only approximate, depending on the actual shape of the distribution and the sample size, and should be used only for providing a *rough* check on obtained values of *SD*. For example, if the range of a distribution is 120 and *SD* is computed to be 9.2 ($R/SD \cong 13$), there is probably something wrong with the calculations. Or, if $R = 24$ and the calculated $SD = 27$, something is wrong since *SD* is greater than *R*. However, if a computed *SD* is such that R/SD is, say, 7.5 for $N = 300$, this is not grounds for suspecting a computational error. The size relationship between *SD* and *R* is useful as a check on gross computational errors only.

Use of SD. We know that *SD* is to be used as a statistic for describing the variability within a distribution and also for differences in variability among distributions. For example, suppose a large sample of concrete cylinders were tested for compressive strength; $\overline{X}$ is reported as 2000 psi and *SD* 60 psi. Given this information about $\overline{X}$ and *SD* and assuming the distribution of cylinder strengths is the familiar bell-shaped one, we would expect almost all strengths to lie between 1800 and 2200 psi with the majority probably falling in the 1900– to 2100-psi range. The distribution would probably look something like the upper curve of Figure 14-4. If *SD* in this example were, say, 200 psi, we might expect some strengths to fall below 1500 psi and some above 2500 psi since the dispersion is much greater, as indicated by *SD*. This distribution would presumably look something like the lower curve of Figure 14-4.

These conclusions certainly are rather vague; the example is intended to show the kinds of things that can be inferred about a distribution of frequencies when *SD* and $\overline{X}$ are known. We will see in the next chapter, on the normal distribution, how when certain assumptions are made, *SD* can be used to obtain much more exact information about the dispersion of values within a distribution.

In Figure 14-4, the values of the variate at 1, 2, and 3 *SD*'s above the mean and 1, 2, and 3 *SD*'s below the mean have been indicated for each curve. Thus, in the upper curve, where $SD = 60$, 2060 psi is 1 *SD* above $\overline{X}$ and 1940 psi is 1 *SD* below $\overline{X}$. Similarly, $\overline{X} + 2\ SD$ is 2120 psi and $\overline{X} - 2\ SD$, 1880 psi; while $\overline{X} + 3\ SD = 2180$ psi and $\overline{X} - 3\ SD = 1820$ psi. In the lower curve, where *SD* is 200, similar points at ± 1, ± 2, and ± 3 *SD*'s have been indicated. It is evident from these graphs that *SD* can be thought of as a unit of measurement along the baseline of a frequency polygon, *measured from the mean*. This visualization of *SD* on the graph of a frequency distribution will play an important part in the next chapter.

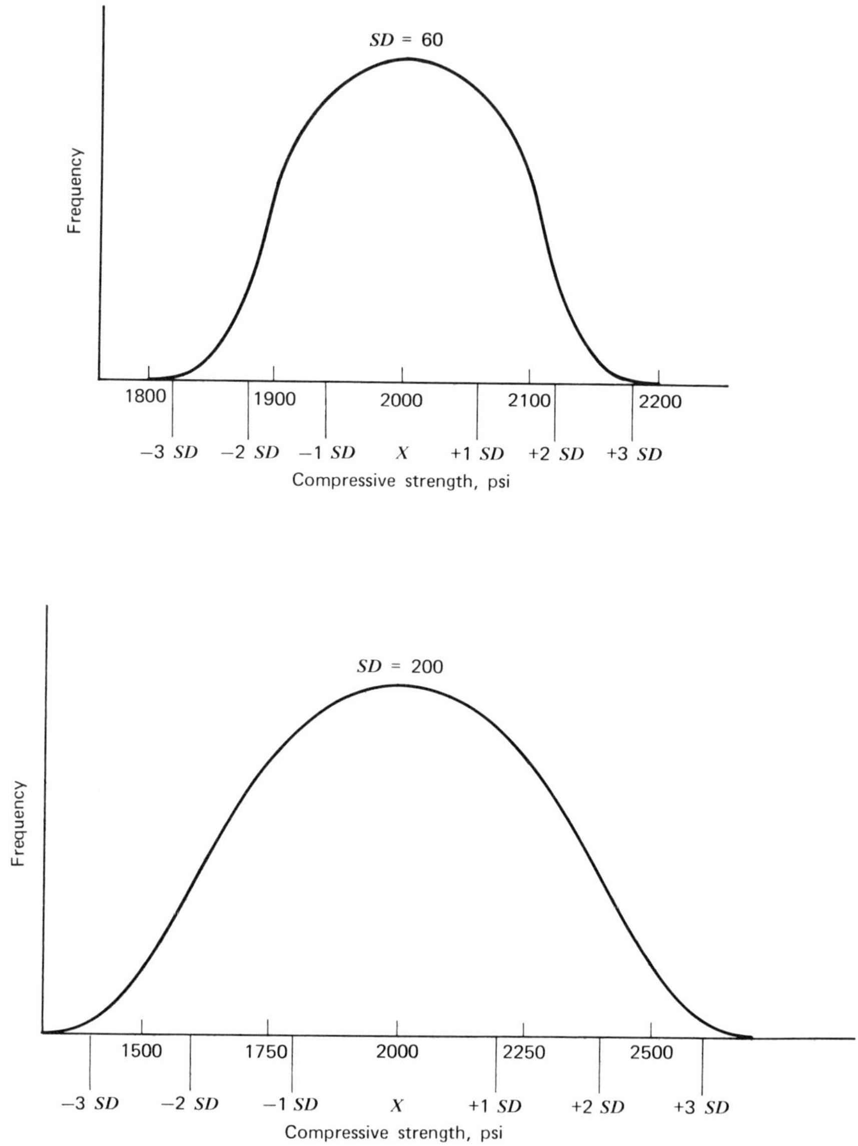

Figure 14-4 Two hypothetical distributions with the same *X* and different values of *SD* (see text).

Coefficient of Variation

An adaption of *SD* is the coefficient of variation, v, obtained by expressing *SD* as a percentage of $\overline{X}$. That is,

$$v = 100\frac{SD}{\overline{X}}$$

Whereas *SD* is measured in the same units as the data, v is a percentage.

The coefficient of variation is sometimes useful when comparing variabilities (*SD*'s) of two or more distributions relative to their means. If, for example, we wish to compare variabilities of 7- and 28-day compressive strengths of concrete, we might get the hypothetical results shown in Table 14-K.

This example shows that whereas the variability of 28-day strength measures was considerably greater as evidenced by the greater *SD*, the relative variability was actually less, as shown by the coefficient of variation, v.

Statistical Presentation of Data

In this chapter we have presented several descriptive statistics, functions of the data that convey some information about the original set of observations. The total information can be presented only by giving all of these original values in the form of an ungrouped frequency distribution. Much of this information, however, is contained in a few functions derived from the data.

No single such function, presented alone or even together with the number, *N*, of observations, is capable of giving much of the total information in the original distribution. For example, much information of value is usually lost by presenting merely the mean and failing to give some measure of dispersion and the number of observations, since quite dissimilar distributions can have identical $\overline{X}$'s. Only by presenting at least two of these functions, together with some statement regarding the sample size, can a fairly accurate description of the distribution be made.

Table 14-K
VARIABILITIES OF HYPOTHETICAL SAMPLES
OF 7- AND 28-DAY CONCRETE STRENGTHS

	$\overline{X}$, psi	*SD*, psi	v, percent
7-day strength	1500	180	12.0
28-day strength	3500	315	9.0

Generally, if a set of measures of a single variable is obtained under *controlled conditions* (Chapter 13), much of this information will remain accessible if X, SD, and N are presented. Why this is so will become clearer after reading the next chapter.

Just what functions of the data should be presented in a particular instance depends on what uses are to be made of the data. Presentation of data or descriptive statistics presumes some intended use either by others or by the author as supporting evidence. The objective is to present that portion of the total information given by the original data that is believed to be essential for the intended use.

If we are interested in the percentage of the total number of observations that have values above, below, or between certain values on the scale of measurement, the essential information can be provided by a tabular grouped distribution of relative frequencies together with a statement of the number of observations, N. But, if N is large and the data were obtained under controlled conditions, the necessary information may be communicated by giving X, SD, and some indication of degree and direction of skewness, as well as N. If conditions under which the data were obtained were not sufficiently controlled, a statement of the maximum and minimum measures may be of value.

If we are interested in the average and variability of the quality of a material, or in the average quality of a material and some measure of the variability of averages for successive samples, or in comparing averages and variabilities of different materials, the $\overline{X}$, SD, and N of each set of measures would provide the necessary information.

If N is quite small, say 10 or less, it is probably better to give R (range) rather than SD. Whenever R is presented, a statement of the sample size, N, should accompany it; this is because the value of R depends to an extent on N, especially when N is small. This dependence is intuitively obvious since the greater the number of items sampled, the more likely it is that extreme values, those at the tails of the distribution, will be included, and, of course, R increases with the addition of extreme values.

Sometimes when a distribution is markedly skewed, it is possible to give a distorted impression of the data when $\overline{X}$ is the only measure of central tendency given, especially when the distribution is sharply peaked. For example, a frequency distribution of annual incomes for a United States town may be heavily skewed due to the presence of a few atypical, extremely high incomes. In such a case, the modal and median incomes might both be around \$10,000, say, while the mean, or average, income might be two or three times that amount. Quoting $\overline{X}$ alone would not be appropriate and could easily mislead an uninformed reader. Instead, what is usually done in the case of highly skewed distributions is to report *Mdn.* in addition to $\overline{X}$—or even all three measures. From these, the direction and amount of skewness can be inferred and proper interpretation of the data made.

The following general recommendations for presenting data are a summary of points made above.

1. Present as a minimum $\overline{X}$, *SD*, and *N*.
2. If *N* is large and it is desired to give information about the shape of the distribution, present also a measure of skewness, or the *Mdn.* and/or *Mo.*, or a grouped frequency distribution (tabular or graphical) if necessary.
3. If the data were not obtained under controlled conditions and it is desired to give information regarding extreme effects, present the maximum and minimum values of the data, in addition to $\overline{X}$, *SD*, and *N*.

CHAPTER 15
NORMAL DISTRIBUTION, CONFIDENCE LIMITS

THE NORMAL DISTRIBUTION CURVE

So far we have learned methods that may be used to *describe* distributions of data. In trying to make a set of data meaningful to someone else, usually the first thing we do is present descriptive statistics, such as the mean and standard deviation. But often this is not enough; commonly, we will want more information from a distribution than can be obtained from descriptive statistics alone. As an illustration, consider the following example.

Suppose strengths of a large series of 6×12-in. concrete cylinders were measured; X was computed to be 4200 psi and SD 510 psi. We might be interested in knowing, for example, what proportion of cylinders have strengths exceeding 3500 psi or, say, what percentage of cylinders have strengths less than 3000 psi. Another such question could be: What strength level is exceeded by 95% of the measures? To answer these and similar questions, we need to know the properties of a particular distribution called the *normal distribution*, one of the most important tools in statistics.

If we had access to the original data for these strength measurements, we could answer the above questions by consulting a tabular ungrouped frequency distribution or a cumulative distribution without using further statistical techniques. The objection to this is that, in almost all cases, we are interested in a larger population from which a sample of data is taken. In the above example, cylinders may have been cast to arrive at some conclusion about strength properties of concrete being produced for some project, in which case the questions that we asked really apply to this manufactured concrete as a whole. For such problems knowledge of the normal distribution, or *normal curve* as it is commonly called, is essential.

The normal curve is shown in Figure 15-1. Before examining some of its characteristics, it should be noted that the normal curve is not a distribution of actual data since it was not obtained by measuring something. It is a *theoretical* distribution, the graph of a mathematical equation, and assumes that the number of cases, N, is infinite. The normal curve is a frequency distribution curve, and the ordinate or Y-axis indicates frequency and the abscissa or X-axis the units of measurement. Since the curve is based on a theoretical population of measures (infinite in number), rather than a sample, the mean and standard deviation are parameters and are symbolized by the Greek letters μ (mu) and σ (sigma), respectively, to distinguish them from the statistics, $\overline{X}$ and SD, which apply to actual obtained data. (See Chapter 13.)

For the normal curve, the units of measurement along the baseline are always expressed in *standard deviation units* as shown in Figure 15-1. The number of such units above or below the mean is designated by z; for example, the point on the normal curve 2 standard deviations below μ corresponds to $z = -2$.

The most obvious characteristic of the normal curve is its shape. It is a symmetrical, bell-shaped curve, rising to a smooth peak in the middle and tapering off symmetrically into tails at either side. Since the normal curve is a theoretical distribution based on an infinite population size (N), these tails will taper indefinitely—always approaching but never quite reaching the X-axis. For practical purposes, the normal curve is usually cut off at three standard deviations above and below the mean ($z = +3$) as shown in Figure 15-1, since, as we shall see, a very small proportion of frequencies falls beyond these values.

Another characteristic of the normal curve is the coincidence of the mean, median, and mode, all three laying at the center of the distribution. This is because the distribution is symmetrical.

We have seen that the normal curve is a particular theoretical distribution that is symmetrical. There are, of course, countless other sym-

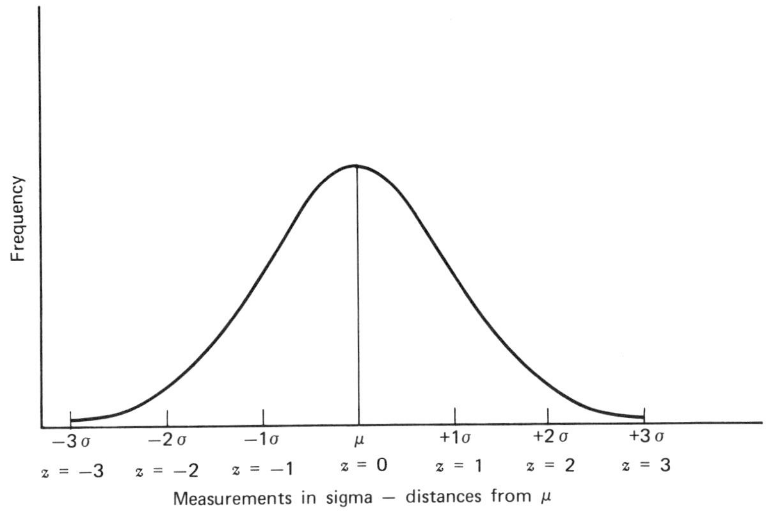

Figure 15-1 The normal distribution curve.

metrical distributions of various shapes and, in fact, there are an unlimited number of bell-shaped curves similar to the normal curve. It is clear that we are emphasizing one specific curve. Before we go on to examine some of the characteristics of this distribution, let us see why this is so.

Since we will always be dealing with distributions of data obtained by measuring something, that is, *empirical* distributions, why is the theoretical normal curve so important? The reason is that many empirical distributions are shaped approximately like the normal curve. Furthermore, when sample sizes are large, these distributions are even closer approximations to the normal curve. In such cases, we assume that the population distribution from which the sample has been drawn, and with which we are ultimately concerned, is an exact normal curve (or, in common terminology, is *normally distributed*).

In particular, the distribution of the net effect of so-called *random* errors that collectively result in unavoidable variations in measures can be expected to be approximately normally distributed. These accidental errors are in contrast to *systematic* errors that arise from causes that act consistently under given circumstances and are considered subject to control. Some of the many factors that might cause unavoidable variations in the strength of concrete cylinders, for example, are changes in climate, changes in aggregate and cement sources, variations in cement manufacture, differences in maximum size of aggregate, and alterations in the water-cement ratio. Some of these factors increase and others decrease the value of an observation being made. It is this combined effect of errors acting in opposite directions that causes variation between strength measures even under carefully controlled conditions, and causes these measures to be approximately normally distributed.

Whenever a set of data is considered to be normally distributed, the normal curve can be used to obtain a great deal of information from the data.

Area under the Normal Curve. Since the normal curve is a particular distribution with a precise configuration, it follows that there is an exact relationship between distance along the baseline and the area under that portion of the normal curve. This can be seen in Figure 15-2. The total area under the curve, that is, the area between the curve and the baseline, is taken as 100%. The figure shows the proportions of the total area under the curve lying between $\bar{X}$ and values of z equal to 1, 2, and 3.

These percentages also represent the same proportions of the total *frequency* because the total area under the curve is really the sum of a large number of imaginary lines dropped from each point on the curve to the baseline, the length of each line being equal to the frequency at that point. Thus, in the normal curve (Fig. 15-2) 34.13% of the total frequency is included between $\bar{X}$ and the point one standard deviation above

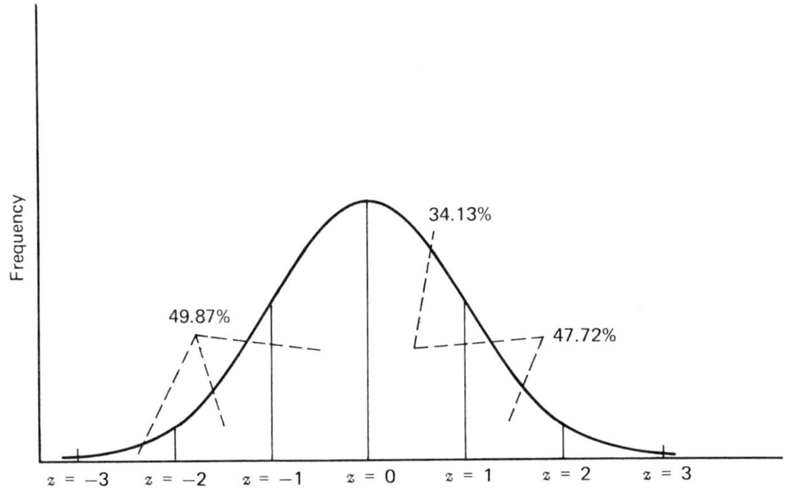

Figure 15–2 Areas under the normal distribution curve.

$\overline{X}$ ($z = +1.00$). Since the normal curve is symmetrical, another 34.13% of the frequency lies between $\overline{X}$ and the point one standard deviation below $\overline{X}$, between $\overline{X}$ and $z = -1.00$. Therefore, 68.26% of all cases lie between $z = -1.00$ and $z = +1.00$. From $\overline{X}$ to the point 2 standard deviations above $\overline{X}$ in the normal distribution, 47.72% of the total frequency is included, so 95.44% falls between $z = -2.00$ and $z = +2.00$. Three standard deviations above and below $\overline{X}$, 6 standard deviations in all, include 99.74% of the total frequency, almost the entire distribution. The remaining 0.26% represents the percent frequency in the two tails more than 3 standard deviations from $\overline{X}$.

These areas are expressed as decimal fractions in Table 15-A, together with areas for certain other selected values of z that are of special significance. The middle column gives the area between X and z and the right column twice this value—the area between $-z$ and $+z$. Complete tables giving areas under the normal curve for all values of z (commonly in steps of 0.01) can be found in any textbook on statistical methods; for our purposes, the values in Table 15-A will suffice.

Example Problems. As a first example of the use of the normal distribution, consider the following problem:

> Measurements of the finished inside diameter of a sample of freshly manufactured concrete pipe were taken; the frequency distribution of these measures was similar to the normal curve with $X = 14.00$ in. and $SD = 0.02$ in. If specification limits are 14.00 ± 0.05 in., what percentage of pipes can be expected to be unacceptable?

What we are being asked for in this problem is the total percentage of pipes whose diameters either exceed 14.05 in. or are less than 13.95 in.;

Table 15-A
SELECTED AREAS UNDER THE NORMAL CURVE

z	Area between $\overline{X}$ and z	Area between $-z$ and $+z$
0	0.0000	0.0000
1.00	0.3413	0.6826
1.96	0.4750	0.9500
2.00	0.4772	0.9544
2.50	0.4938	0.9876
2.58	0.4950	0.9900
3.00	0.4987	0.9974

that is, the percentage of pipes with diameters differing from the mean by more than 0.05 in. Since the pipe diameters can be assumed to be normally distributed, we could answer this question using a table of normal curve areas if we knew the value of z corresponding to a deviation (x) of 0.05 in., which would express this deviation in *SD* units.

To convert x to a corresponding z value (in order to use the normal curve), we merely need to divide x by *SD*. Since *SD* is given as 0.02 in., $z=2.5$ (0.05 in./0.02 in.). Thus, a deviation of 0.05 in. in this distribution lies at a point 2.5 *SD*'s from $\overline{X}$ since 1 $SD=0.02$ in.

From Table 15-A, we see that 98.76% (0.9876) of the total frequency lies *between* $z=-2.50$ and $z=+2.50$. Since the problem asks for the percentage frequency outside these limits, we are interested in the frequency *remaining* from the total and have to subtract the above value from 100% (1.0000). This gives 1.24% as the expected percentage of unacceptable pipes. The solution is illustrated in Figure 15-3, where the shaded tails of the curve are the regions corresponding to unacceptable pipes.

As another example, we will change the above problem slightly: Suppose the distribution of diameter measurements was shaped as before, but $\overline{X}$ was found to be 14.01 in. rather than 14.00 in. (*SD* is still 0.02 in.).

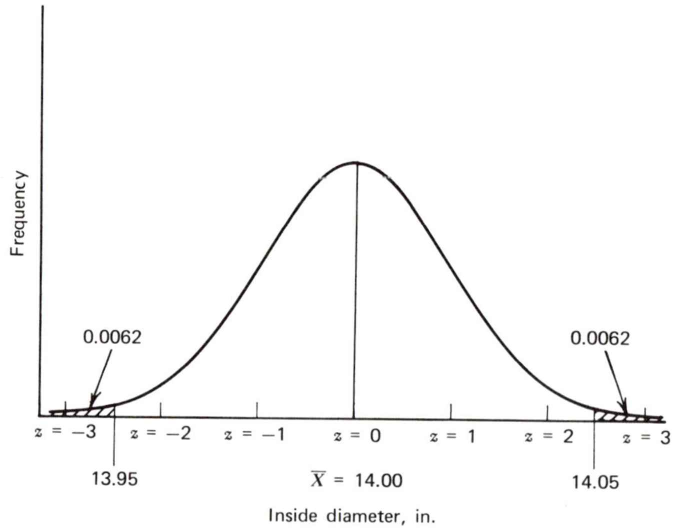

Figure 15-3 Illustrating percentage unacceptable pipe for example problem where $\overline{X}=14.00$ in.

Let us see how Table 15-A can be used to find the percentage of unacceptable pipe in this case.

The specification limits are the same as before—14.05 and 13.95 in. Hence, since $\bar{X}=14.01$, the critical values of z above and below the mean will be different: the former corresponding to a deviation of 0.04 in. $(14.05-14.01)$ and the latter to a deviation of 0.06 in. $(14.01-13.95)$. The z value corresponding to an x of 0.04 in. can be found by dividing x by *SD*—an operation that converts the deviation to *SD* units, as we saw in the first example. This gives 2.00 for z since *SD* is 0.02 in. Similarly, z for $x=0.06$ in. is 3.00. In this problem, then, we are looking for a total percent frequency equal to the sum of the percentage of frequencies above $z=+2.00$ and the percentage of frequencies below $z=-3.00$.

We are now in a position to use Table 15-A. In the middle column, we see that 47.72% of the frequencies lie between $\bar{X}$ and $z=+2.00$, while 49.87% lie between $\bar{X}$ and $z=-3.00$. Therefore, 97.59% $(47.72+49.87)$ of the total frequency falls between $z=-3.00$ and $z=+2.00$. Subtracting this from 100% gives the percent frequency lying outside these values, 2.41%. This, then, is the expected percentage of unacceptable pipes. This is illustrated in Figure 15-4, where once again the shaded tails represent the regions of unacceptable pipes.

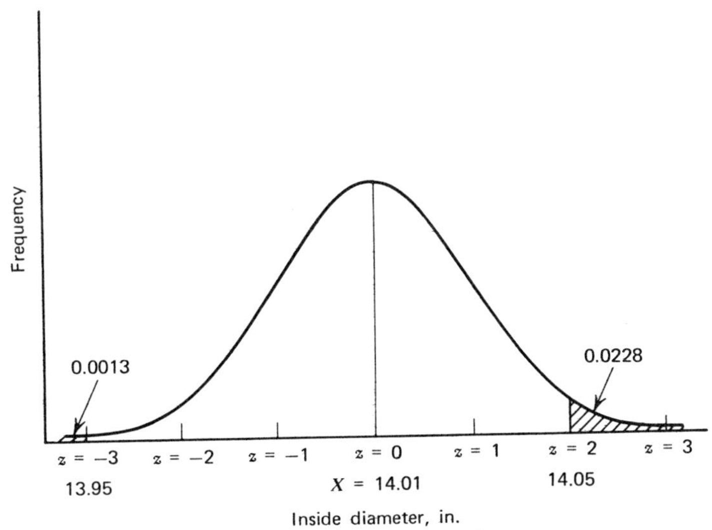

Figure 15–4 Illustrating percentage unacceptable pipe for example problem where $\bar{X}=14.01$ in.

As a third example, consider the following problem:

Compressive strengths of a large number of concrete cylinders were measured; $\overline{X}$ was found to be 3240 psi and *SD* 455 psi. Assuming a normal distribution, find the "minimum" cylinder strength that can be used for design purposes, accepting a definite risk that 2½% of test cylinders will have strengths less than this minimum.

In this problem, we are asked to determine the strength level below which exactly 2.5% of the test strengths can be expected to fall. This means that 47.5% of the strengths will lie between this value and *X* since half the frequencies of the normal distribution lie below the mean. (See Fig. 15-5.)

Entering the middle column of Table 15-A, we see that 47.5% of the total area is included between $\overline{X}$ and a value of $z=1.96$. Hence, the cylinder strength below which 2.5% of the frequencies can be expected to lie is that strength value 1.96 *SD*'s below $\overline{X}$ in the distribution. Since $SD=455$ psi, a deviation (x) of 1.96 *SD*'s is 892 psi (1.96×455), and the required strength is $3240-892$, or 2348 psi ($\overline{X}-x$). The solution is shown in Figure 15-5.

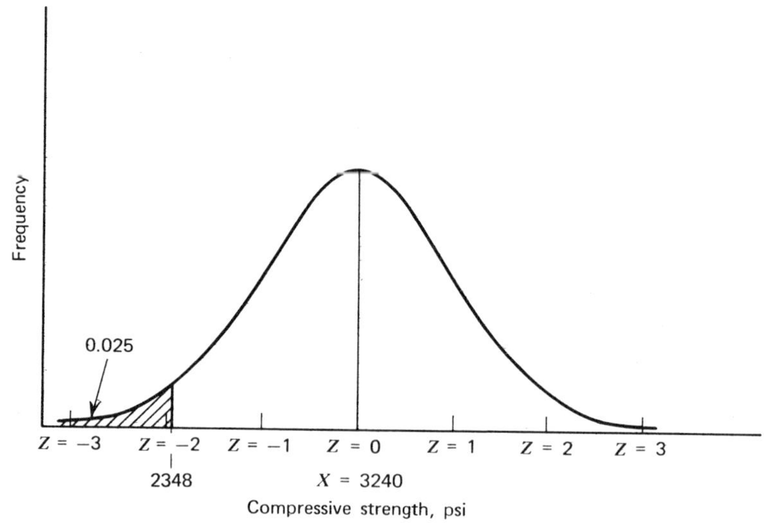

Figure 15–5 Illustrating "minimum" concrete cylinder strength for third example problem.

This problem is the inverse of the two previous problems. In the examples concerning concrete pipe, we were given certain critical values of X (14.05 and 13.95 in.) and required to convert the corresponding deviations (x's) to z values in order to use the table of normal curve areas to obtain certain frequencies (proportions of unacceptable pipe). In this last example, involving strengths of concrete test cylinders, we were given a percent frequency ($2\frac{1}{2}\%$ in the lower tail) and had to use the table of normal curve areas to obtain a corresponding value of z, after which z was converted to a deviation, x, which finally determined the required value of X (minimum strength for design purposes). This is pictured in the following diagram where

$$X \longleftrightarrow x \longleftrightarrow z \xleftrightarrow{\text{Table 15-A}} \%f$$

the two-headed arrows indicate taking the steps either from left to right or from right to left. In either case use of a table of normal curve areas is essential as a link between z and $\%f$, indicated in the diagram by reference to Table 15-A.

CONFIDENCE INTERVALS OF A MEAN

From Table 15-A, it can be seen that 95% of the total frequency of a normal distribution lies between the values 1.96 standard deviations below the mean and 1.96 standard deviations above the mean. It follows that the chances are 95 in 100 that a sample observation will fall within the range $\mu \pm 1.96\sigma$ when observations are normally distributed. Stated another way, we can say that the *probability* of an observation falling within the range $\mu \pm 1.96\sigma$ is 0.95 (or 95%). Similarly, the chances are 99 out of 100 (the probability is 0.99) that an observation will fall within the range $\mu \pm 2.58\sigma$. These statements of probability are the kind used when we are not able to make an assertion about the exact value of a population characteristic (parameter) and are instead interested in establishing a probable range of values.

When we make assertions about a population parameter falling within a specific range of values with a certain probability of being correct, a special terminology is employed. The associated degree of confidence, the probability of being correct, is called the *confidence level*. The corresponding range is called the *confidence interval*, and the limits of the confidence interval are the *confidence limits*.

Estimating a Population Mean from a Sample

Confidence intervals are usually employed to express the uncertainty associated with an estimate of some population parameter. The concept is of considerable importance as it indicates the reliability of our estimate: the narrower the confidence interval at a specific confidence level, the more precise the estimate.

In this section, we will discuss the establishment of suitable confidence limits for a population mean (μ) when such a mean is estimated on the basis of a single sample taken from the population.

The purpose of measuring a sample from a population is to get data that will allow us to make statements about characteristics of that population. The first statistic we would probably compute from this data is $\overline{X}$. We might then ask how close this sample mean is to the true population mean (μ), the mean we would get if we measured the entire population. As we might expect, $\overline{X}$ for a single sample will rarely be precisely the same as μ. In fact, if we took a series of samples from a population with a known μ, we would find that few, if any, of the sample $\overline{X}$'s would be identical to μ. The fact that sample $\overline{X}$'s fluctuate from the population mean is known as *sampling error* of $\overline{X}$'s.

When it is necessary to estimate μ from a sample, what is commonly done is to use the data to make an assertion, together with a specified confidence level, about a confidence interval within which μ might be expected to fall, rather than assign a precise value to μ. The confidence levels most frequently used are the 95% and 99% confidence levels and the corresponding intervals are called the 95% confidence interval and 99% confidence interval.

Confidence intervals for μ are usually presented in the following form:

$$\overline{X} \pm a(SD)$$

where $\overline{X}$ and SD apply to the sample, and a is a factor whose value is such that the resulting confidence limits enclose an interval that can be expected to contain μ with a degree of confidence indicated by a specified confidence level. This value of a depends on the sample size, N, as well as the specified confidence level.

In order to compute confidence limits for an estimated population mean, μ, it is necessary to determine the value of a in the above expression for a desired confidence level and the size, N, of the sample. In order to understand how this is done, let us first see how the normal curve is relevant to this problem.

Sampling Distribution of Sample Means. Suppose we take successive random samples from a large population. We specify *random* samples because it is important that the samples be representative of the total population. A random sample, you will remember is one in which each item in the population has had the same chance of being chosen for the sample as every other item. If we take a large number of such samples, each of size N, and measure them on some characteristic, we can compute $\overline{X}$ for each of these samples.

Let us now consider the frequency distribution of these $\overline{X}$'s. If we group the $\overline{X}$'s and place them along the baseline (X-axis), and plot the frequencies along the Y-axis, we then have a grouped frequency distribution of sample $\overline{X}$'s. Such a distribution is called the *sampling distribution* of $\overline{X}$. What is the nature of this distribution? Assuming that the samples were relatively large, we would find that the X's would fall into an approximate normal distribution. Furthermore, the mean of this distribution—the mean of all the sample $\overline{X}$'s—would be the best single estimate of the population mean, μ.

Now, if we knew SD of this sampling distribution, we could make statements about the likelihood of subsequent sample $\overline{X}$'s falling within specified intervals of the distribution, since this distribution follows the normal curve. Thus, just as we did before for a sample observation, we can say, for example, that the chances are 95 out of 100 that any $\overline{X}$ will fall between 1.96 SD's above and below the mean of the distribution.

The above procedure is one way of providing confidence limits for a population mean. But it might seem that if we were going to obtain a large number of samples in order to get the SD of the sampling distribution, we might as well measure the entire population, if possible, thereby giving us μ and avoiding problems of sampling error. At any rate a method of repeated sampling is tedious, and fortunately it is unnecessary for the problem of establishing confidence limits for μ. This is because there is a technique for estimating SD of a distribution of sample means from the data of a *single sample*, which brings us back to the original problem.

Estimating the Standard Deviation of the Sampling Distribution. Statisticians have shown that there is an exact relationship between the standard deviation of a distribution of sample $\overline{X}$'s and the standard deviation of the population from which the samples were taken and the size (N) of each of these samples. This relationship is expressed by the formula

$$SD_X = \frac{\sigma}{\sqrt{N}}$$

where SD_X is the standard deviation of the sample $\bar{X}$'s and σ is the true population standard deviation (of all the X's).

In the problem at hand we are working with a single sample and do not have access to the population standard deviation, σ. What we can do, however, is use SD of the sample to *estimate* σ. This technique requires a brief explanation itself.

If we wished to estimate a population mean (μ) from a single random sample, the best guess would be the sample mean, $\bar{X}$. This provides what is called an *unbiased* estimate. If we want to estimate a population standard deviation (σ), however, the sample SD gives a *biased* estimate because SD tends to be slightly smaller than σ, the discrepancy depending on the sample size (N). The reason why the sample SD is smaller than the population value is that with fewer and fewer items included in a sample, it becomes less and less likely that extremely deviant values contained in the population will be included within the sample. These extreme cases contribute relatively large squared deviations to the population standard deviation. This greater possibility of including extreme values when samples are made larger was also mentioned in Chapter 14 when discussing the relation between the range (R) and N.

Returning to the relationship $SD_X = \sigma / \sqrt{N}$ where we wish to estimate σ using the SD of the sample, we can compensate for this underestimation of σ by SD by dividing SD by the square root of $N-3$ instead of N provided N is not too small. In other words, dividing SD by $\sqrt{N-3}$ gives a close (although for certain technical reasons still slightly biased) estimate of $6/ \sqrt{N}$, the factor $\sqrt{N}/ \sqrt{N-3}$ tending to compensate for the underestimation of σ by SD. The estimate of the standard deviation of sample means is thus found by using the formula

$$SD_X = \frac{SD}{\sqrt{N-3}}$$

All we have to do, then, to estimate $SD_{\bar{X}}$ is to compute SD of the sample measures and divide this SD by the square root of 3 less than the size of the sample. Let us state once more what this estimate represents. It is an estimate, based on the data from a single sample, of the standard deviation of a distribution of a large number of $\bar{X}$'s of successive random samples of the same size as the single sample. This estimate in turn allows us to make statements about the interval in which the true mean (the population mean) lies at any desired confidence level since the distribution of these $\overline{X}$'s is normal (follows the normal curve).

Formulas for Confidence Limits

Let us now return to the specific problem of establishing confidence limits for a mean. We pointed out before that such limits are indicated in the form

$$\mu = \overline{X} \pm a(SD)$$

where a has to be determined and depends on the confidence level and N. We also just saw that an estimate of the standard deviation of the distribution of a large number of sample means can be obtained by dividing SD by $\sqrt{N-3}$.

Let us see what the value of a in the above expression must be when establishing the 95% confidence interval for μ. We know that for a normal curve the probability is 0.95 that a value of the variate will fall in the interval between plus and minus 1.96 standard deviation's from the mean. We can make a similar probability statement about where μ will fall in the distribution of sample means (a normal distribution), namely: The probability is 0.95 that μ will fall in the interval between plus and minus 1.96 $SD_{\overline{X}}$'s from the mean, where $SD_{\overline{X}}$ is the standard deviation of the sample means that we have estimated as $SD/\sqrt{N-3}$. Since $\overline{X}$ is the best *single* estimate of μ from a single sample, computation of the limits $\overline{X}+1.96(SD/\sqrt{N-3})$ and $\overline{X}-1.96(SD/\sqrt{N-3})$ establishes the 95% confidence interval. Thus, the value of a for a 95% confidence level is $1.96/\sqrt{N-3}$ and the formula for 95% confidence limits becomes

$$\mu = \overline{X} \pm \frac{196\ SD}{\sqrt{N-3}}$$

In setting up this 95% confidence interval, the chances are 95 out of 100 that μ does indeed lie within this range of values.

Similar reasoning leads to a formula for 99% confidence limits of a mean:

$$\mu = X \pm \frac{2.58\ SD}{\sqrt{N-3}}$$

That is, the sample $\overline{X}$ plus and minus 2.58 times $SD\ /\ \sqrt{N-3}$ establishes the 99% confidence interval for μ.

We can, of course, set up confidence intervals at levels other than 95 or 99%. Ninety-percent confidence intervals are sometimes used, for example, and if we wanted an especially high degree of probability, we might wish to establish 99.9 or even 99.99% intervals. Generally, however, 95 and 99% intervals have proven most useful, and these levels are the ones usually encountered in statistical presentations.

Example. The following example illustrates the procedure for computing confidence limits:

> Tensile strength tests were made on 124 steel cables chosen at random from a certain mill. X and SD of the sample were computed as 192.1 and 13.2, measured in thousands of pounds per square in. (kip/in.^{2}). Establish 95 and 99% confidence intervals for the mean of the population.

The formula for 95% confidence limits is

$$\mu = X \pm \frac{1.96\ SD}{\sqrt{N-3}}$$

In this problem X is 192.1, SD 13.2, and N 124. Therefore

$$\frac{SD}{\sqrt{N-3}} = \frac{13.2}{\sqrt{121}} = \frac{13.2}{11} = 1.2$$

and

$$\frac{1.96\ SD}{\sqrt{N-3}} = 1.96 \times 1.2 = 2.4$$

The 95% confidence interval is thus

$$192.1 \pm 2.4 \text{ kip/in.}^2$$

and the 95% confidence limits are 189.7 and 194.5.

The 99% confidence interval is given by

$$\overline{X} \pm \frac{2.58\ SD}{\sqrt{N-3}}$$

Since

$$\frac{2.58\ SD}{\sqrt{N-3}} = 2.58 \times 1.2 = 3.1$$

the 99% confidence interval is

$$192.1 \pm 3.1 \text{ kip/in.}^2$$

and the corresponding limits are 189.0 and 195.2.

Confidence Limits for Small Samples. In discussing the sampling distribution of X, we said that this distribution would be approximately normal provided the size of each sample, N, was relatively large. The fact that this sampling distribution follows the normal curve allows us to set up confidence intervals based on areas under the normal curve, reflected in the use of the constants 1.96 and 2.58 in the formulas for 95 and 99% confidence limits, respectively.

With small samples, however, the distribution of $\overline{X}$'s, while symmetrical, departs from normality, the departure being greater for smaller samples. When we have measured only a small sample—say 25 cases or less—and wish to set up a confidence interval, we therefore cannot use the formulas for confidence limits given above. If we did, the resulting confidence intervals would be unjustifiably small.

What we can do, however, is use the same basic formula for confidence limits but change the value of a according to a special table (Table 15-B). The fundamental expression for confidence limits of a mean, you will remember, is

$$\mu = \overline{X} \pm a(SD)$$

where a depends on the sample size and the confidence level. For larger samples (N over 25) we found that $a = 1.96/\sqrt{N-3}$ for 95% confidence limits and $a = 2.58/\sqrt{N-3}$ for 99% confidence limits. For small samples all we have to do is use Table 15-B to find the value of a for the N of the sample and the confidence limits (95 or 99%) we are interested in.

To illustrate the procedure, we will establish 95 and 99% confidence intervals for the mean for the example worked before, assuming, however, that only 20 steel cables were tested. In Table 15-B values for a for $N = 20$ are given as 0.480 for 95% confidence limits and 0.656 for 99% confidence limits. Therefore, the 95% confidence interval is given by

$$192.1 \pm 0.480 \times 13.2 = 192.1 \pm 6.3 \text{ kip/in.}^2$$

and the 99% confidence interval by

$$192.1 \pm 0.656 \times 13.2 = 192.1 \pm 8.7 \text{ kip/in.}^2$$

Presentation of Confidence Limits

When giving confidence limits, it is essential that the confidence level be clearly indicated. Since the width of the interval depends on the size of the sample from which the confidence limits have been computed, it is also necessary to indicate N.

Table 15-B
FACTORS FOR CALCULATING 95 AND 99% CONFIDENCE LIMITS FOR MEANS BASED ON SMALL SAMPLES [confidence limits $= X \pm a(SD)$]*

Sample Size N	Value of a	
	95% Confidence Limits	99% Confidence Limits
4	1.837	3.372
5	1.388	2.302
6	1.150	1.803
7	0.999	1.513
8	0.894	1.322
9	0.815	1.186
10	0.754	1.083
11	0.705	1.002
12	0.664	0.936
13	0.629	0.882
14	0.599	0.835
15	0.573	0.796
16	0.550	0.761
17	0.530	0.730
18	0.512	0.703
19	0.495	0.678
20	0.480	0.656
21	0.466	0.636
22	0.454	0.618
23	0.442	0.601
24	0.431	0.585
25	0.421	0.571

* From *ASTM Manual on Quality Control of Material* (1951), page 43.

One simple and effective way of doing this is the following:

$$\mu = 192.1 \pm 6.3 \text{ kip/in.}^2 (P = 0.95, N = 20)$$

P in such statements refers to the probability level.

As a final word, it should be stressed that nothing is proved about the value of μ by setting up confidence intervals. We must realize that μ could lie outside of these limits; there is no way of proving whether it does or does not, short of measuring the entire population. We simply have a certain degree of confidence, based on probabilities, that μ lies within the confidence interval.

GLOSSARY

GLOSSARY

act of God. An event that occurs through the irrestible force of nature such as flood or tornado.

bimodal. A distribution with two humps or peaks in the frequency distribution.

breach of contract. The breaking of a contract by the failure of at least one party to fulfill its obligations under a contract.

class interval. Each of the equal-sized groups in a grouped frequency distribution.

confidence level. The probability of being correct.

contract. The subject matter of an agreement to which the participating parties give their consent.

contract law. The legal process for interpreting and settling any differences in written or unwritten agreements.

corporation. An artificial entity owned by one or more persons and created and maintained in accordance with the laws of the state.

criminal law. Certain actions that violate the general rights of the entire community would fall under the jurisdiction of criminal law.

critical path method (CPM). A method of planning for job activity that shows the slowest jobs and the order in which things need to be done before other things can begin. One can then devote major emphases to that sequence or path of jobs.

cumulative frequency distribution. The total frequency below or above a certain value.

data. Information, facts.

defendant. The one accused of some violation.

empirical. Measured.

frequency distribution. An arrangement of data that indicates the frequency of each possible measurement.

kurtosis. The degree of peakedness in a bell-shaped distribution curve.

law of agency. The legal requirements and restrictions applying to the actions of any substitute for the primary parties.

law of torts. Laws applying to incidents

of personal injury whether they involve a workman on the construction project or an innocent bystander.

liability. Being legally bound to deliver agreed upon performance or specified compensation for failure to perform or being legally bound to make good any loss or damage.

lien. A claim on the property of another as a security against the payment of a just debt.

mean. Commonly known as the average.

mean deviation. The sum of all deviations above this point exactly equal the sum of the deviations below.

median. The 50th percentile or the middle value with an equal number of distribution points above and an equal number below.

mode. The value that occurs most frequently.

normally distributed. An exact, normal curve.

origin. The intersection of the *X*-axis and the *Y*-axis.

overhead. The expenses that cannot be charged directly to one job, but rather the basic costs of doing business.

parameter. The numerical characteristics of populations.

partnership. An association of two or more persons to carry on as co-owners a business for profit.

percentile. The value of the variate below which the given percentage of the total number of cases lie.

plaintiff. One who claims in court that his rights have been violated.

population. The class about which conclusions are to be drawn.

range (R). The simplest expression of variability.

reasonably prudent man. An imprecise term referring to a hypothetical creature possessed of the exact degree of skill and wisdom to avoid all danger.

relative frequency. The actual frequency divided by the total number of cases.

sampling. Partial representation of the population more efficiently studied than the whole.

sampling distribution. Grouped frequency distribution.

skewed frequency distribution. Any asymmetrical frequency distribution.

sole proprietorship. The ownership of an unincorporated business by one person.

statistics. The numerical characteristics of the data.

suit. Action to secure justice in a court of law; an attempt to recover a right or claim through legal action.

turnkey construction. A team construction method designed to provide clients with more cost and time controls by having designers and builders working together in the design stage of a project.

variability. The dispersion of a measure.

vis major. The result of a human unrest such as a war or national strike.

X-axis. The horizontal baseline of a graph or abscissa.

Y-axis. The vertical line or ordinate.

INDEX